REMOTE SENSING APPLICATIONS IN BRACKISH WATER FISHERIES

About the Authors

Binod C Agrawal, M.S., Ph.D. (Wisconsin), is currently Director, TALEEM Research Foundation and involved in promoting open and distance learning and setting up *Univesity in the Sky*. Until recently he was with Space Applications Centre, Indian Space Research Organisation (ISRO), Ahmedabad where he worked over two decades and pioneered use of qualitative methods for communication research during the world famous Satellite Instructional Television Experiment (SITE). The outcome of the research has been documented in the book entitled *Anthropological Methods for Communication Research* (Agrawal ed. 1985). Dr. Agrawal is founder Director of Mudra Institute of Communication, Ahmedabad (MICA) where qualitative research technique were used for developing health related communication messages.

His publications include over a dozen books on television, computer technology, and anthropological studies. Dr. Agrawal represented India in several international communication research forums.

Currently he is also President of the National Council for Development Communication, India.

Arbind Sinha with post graduate degree in Anthropology and Ph.D. in Sociology worked for more than two decades with Space Applications Centre, ISRO, Ahmedabad. He has wide experience of working with issues related to applications of science and technology for rural development. He was involved with all major social studies on satellite applications in the country including SITE, Kheda, interactive tele-teaching and Jhabua Development Communication Project. His major involvement has been in the field of development communication, disaster communication, and health communication.

Dr. Sinha is a Professor of TALEEM Research Foundation, Ahmedabad and has been, presently, deputed to work as Communication Advisor to the Royal Danish Embassy, New Delhi for their programme on Danish Assistance to the National Programme for Control of Blindness.

He contributed one book, co-edited two books and more than dozen of papers in books and journals of national and international repute.

Sham Behari Sharma, a Masters in Social Welfare from Agra University did his diploma in Journalism. He started his career with National Council of Applied Economic Research and moved over the NCERT and to Delhi Agricultural Project when he studied social dimension of agricultural telecast. Having worked extensively in SITE and post-SITE projects, he is presently working with Development and Educational Communication Unit (DECU), ISRO, Ahmedabad for almost last more than a decade. He has been the interface between DECU and the user organisations for most of the satellite based interactive teleteaching experiments, conducted in India during 1990s.

Some of his papers were published in Communicator.

REMOTE SENSING APPLICATIONS IN BRACKISH WATER FISHERIES

A Socio-Economic Study

Binod C. Agrawal
Arbind Sinha
Sham Behari Sharma

CONCEPT PUBLISHING COMPANY, NEW DELHI-110059

ISBN-81-7022-934-0

First Published 2002

Published and Printed by
Ashok Kumar Mittal
Concept Publishing Company
A/15-16, Commercial Block, Mohan Garden,
NEW DELHI-110059
Phones : 5351460, 5351794
Fax : 091-11-5357103
Email : publishing@conceptpub.com

PREFACE

Keeping in view the broad objectives and programmes of ISRO, from time to time we have been discussing to examine social applications of space technology. The present project was initiated to achieve this goal in a concrete manner where remote sensing techniques have been used to locate Brackish Water Fisheries sites. The initial proposal was discussed with a visiting team of Food and Agricultural Organization (FAO), Rome. The visiting team met Social Research Group and the Scientists of Remote Sensing Area, particularly Dr. Baldev Sahai, Group Director, Remote Sensing Applications Group. As a result of this informal initial discussion, Ministry of Agriculture, Government of India requested us to initiate this project. Accordingly, a two part study was taken up. Remote Sensing Applications Group completed the mapping of Andhra Pradesh, Orissa and West Bengal. On our side, we took up a two phase study. In the first phase, based on the secondary data, an effort was made to evolve selection criteria for locating brackish water fisheries sites in these states. In the meantime, FAO mission team in consultation with the Ministry of Agriculture and respective State Governments identified brackish water sites. Hence, this phase of work was discontinued. In the second phase, on the request of Ministry of Agriculture, six sites were selected for socio-economic study. This report is a result of second phase study.

In this study, Mr. Mario Pedini of FAO was most helpful from the very beginning leading of its successful conclusion. We take this opportunity to thank Mr. Mario Pedini for his continual assistance, support and encouragement in conducting this study. We take this opportunity to thank Mr. Godbole of FAO whose assistance was very valuable during the discussion phase at Rome. Thanks are also due to the FAO sociologist, Dr. Alice Carloni. We take this opportunity to thank Ministry of Agriculture officials especially Mr. V. Sampath who provided invaluable logistic support for the project.

In the initial stages of the project, valuable assistance was provided by Dr. B.N. Vishwanath, Ms. Anjana Vyas, both from the School of Planning, Ahmedabad and Professor U.K. Srivastava of Indian Institute of

Management, Ahmedabad, we are grateful to all of them for their expert assistance.

The moving force in drawing us into this project is Mr. Pramod P. Kale, Director, SAC, who took personal interest in forging the collaboration between Remote Sensing Scientists and us. Also, he continuously encouraged and monitored the project. Our sincere thanks to him. Dr. George Joseph, Director, SAC, Dr. Baldev Sahai, Dr. S. Naik and their team worked with us and answered innumerable queries and provided valuable thematic maps. We are very grateful to all of them.

Within DECU, Mr. K.S. Karnik, Director, took special interest in the project and provided all facilities for its timely completion. The entire team is grateful to him for his support. Our colleague in the Social Research Group Mr. S.R. Joshi assisted in the completion of the project. Mr. Sham Behari Sharma acted as Field Coordinator for data collection alongwith Dr. Arbind Sinha. Ms. Madhavi Joshi and Mr. Bhaskar Patel joined the team at the time of analysis and report writing. Their timely untiring assistance was greatly appreciated. In the first phase of the study, Mr. Ramchandrudu and Mr. N. S. Reddy were involved. We appreciate the help provided by them.

The analysis and data processing were done with the assistance of Mr. Daxesh Shah. Word processing was a joint responsibility of Mrs. Omana Raghavan, Ms. Champa Gajera, Mr. M. Damodaran and Mr. Suresh Shah. Sincere thanks for their painstaking efforts. Finally, Mr. K.S. Vaghela deserves mention for his logistic support.

Authors

ACKNOWLEDGEMENTS

Space Applications Centre, Ahmedabad
P. P. Kale, Director
George Joseph, Associate Director
Baldev Sahai, Group Director, Remote Sensing Applications Group
S. Naik and the Remote Sensing Team.

Development and Educational Communication Unit, Ahmedabad
K. S. Karnik, Director
S. R. Joshi
Arbind Sinha
Sham Behari Sharma
Bhaskar Patel
Madhavi Joshi
Daxesh Shah
Omana Raghavan
M. Damodaran
Champa Gajera
Suresh Shah
K. S. Vaghela

EXECUTIVE SUMMARY

Socio-Economic Aspects of Brackish Water Fisheries in India

The Objectives of the Study

— Provide socio-economic profile of the potential beneficiaries and define the criteria for the selection of the beneficiaries.
— Suggest size of the brackish water shrimp farms and ponds.
— Recommend management system for organizing individual brackish water shrimp farms and ponds.
— Techno-managerial support required to run the individual farms and ponds.
— Study how the benefits/losses can be shared by the individual, ommunity and co-operative or even states.

Methods of Research Study

— Using exploratory research design, Quick and Useful Evaluative Research Investigation (QUERI) technique in combination with Rapid Rural Appraisal (RRA) techniques were employed for six brackish water fisheries sites, namely Dadanpatra and Digha in West Bengal; Jagatjore and Panaspada in Orissa, and P.T. Palem and C.G. Palem in Andhra Pradesh.
— Five member social science research team, on an average, visited nine villages, for this study within the radius of five km of each brackish water site.
— In each village, group discussions were held with a group of 20-30 persons about the proposed brackish water fisheries project.
— Village population, caste composition and household occupation were estimated using triangulation technique in each village.
— On an average 53 individuals purposively selected from each brackish water site were interviewed, having 58.2% male and 41.8% female respondents.
— Case studies on co-operatives were conducted and observations were made about prawn cultivation and other fishing activities.

Results and Recommendations

— Nine out of ten households in the brackish water sites are poor. Except few almost all practice fishing regardless of their religion, caste or occupation in which men, women and children are involved.

— Potential beneficiaries are aware and some what knowledgeable about brackish water prawn cultivation and its economic benefits. Even some are cultivating prawn in their ponds.

— In order to spread the benefits of the project, it is recommended that the beneficiaries should be selected from among those who have not been helped so far under any poverty alleviation programmes. First opportunity should be given to the poorest of the poor from fishermen caste followed by Scheduled Castes/Scheduled Tribes.

— Three levels of management and organizational structure has been proposed. At the lowest level, the individual households will be responsible for the mangement of their 0.5 hac. ponds. These individuals will form primary fisheries co-operatives. These primary fisheries co-operatives will form a farm level co-operative to manage the affairs of the individual farms. At the third level the management of the farm should be handled by Board of Directors on the pattern of Indian Farmers Fertilizers Co-operatives Ltd. (IFFCO). In this way the individual farm will be an industrial unit.

— Training has been considered as an essential element of the project and the same has been suggested at all levels including the individual farmers.

— Varying degree of infrastructural facilities are available for semi-intensive shrimp farming least being in Jagatjore (Orissa) and most in Digha (West Bengal).

— Idea of no subsidy seems to have not bothered the potential beneficiaries.

— Two kinds of views were expressed about the management and organization of the shrimp farms : (a) large size farms to be managed by private companies and (b) co-operatives of individually managed small ponds with common infrastructure and services of the farms.

— There was greater support both from potential beneficiaries and fisheries officials for semi-intensive brackish water shrimp cultivation co-operatives.

— Giving individual pond ownership/right has been asked by nine out of ten potential beneficiaries. Similar opinion have been expressed by the fisheries officials. This seems to be both psychological and cultural needs of the beneficiaries.

— Keeping in view that a large majority of the poor should benefit from the project and to satisfy the cultural needs of the potential beneficiaries, it is recommended that half hactre ponds should be constructed.

— Since women's participation in semi-intensive brackish water shrimp farming is seen as important for effective operation and management of the individual ponds, it is suggested that the pond ownership/right be given to couple.

CONTENTS

LIST OF TABLES

LIST OF GRAPHS

LIST OF MAPS

Map 1 : Brackish Water and Inland Fisheries Project
Project Site Locations

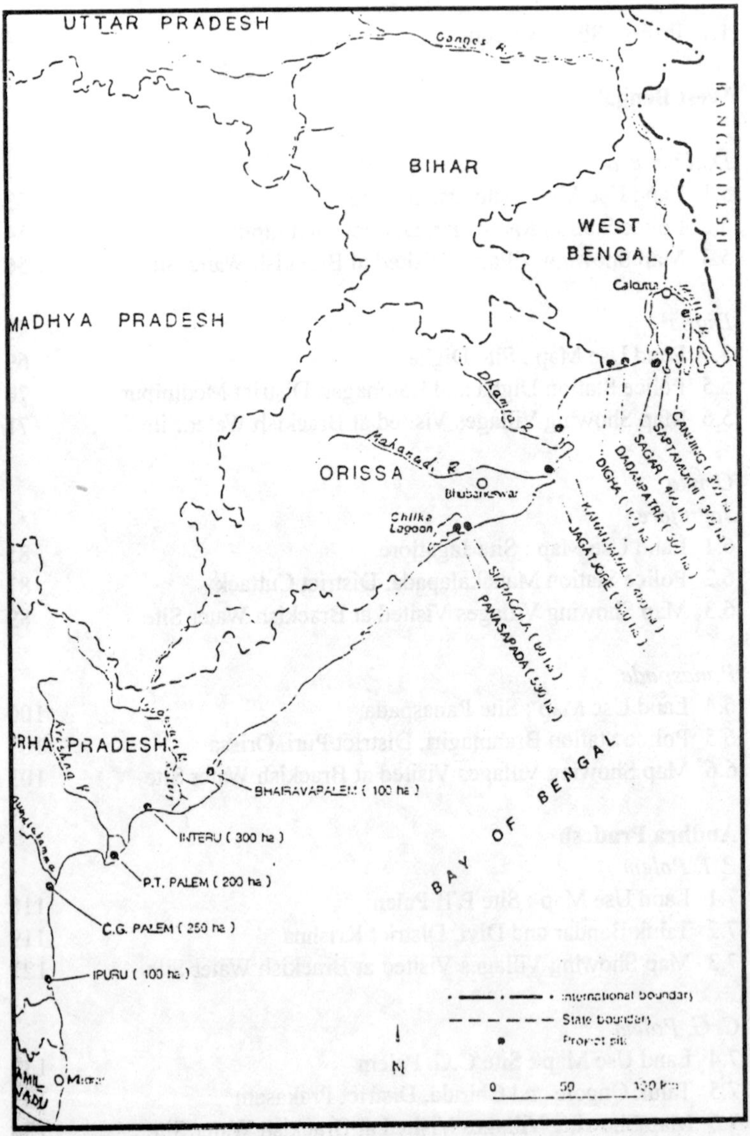

Source : FAO Report.

1

INTRODUCTION

Background

This report emerged as a result of an empirical research study conducted in six selected Brackish Water Fisheries sites identified in a Draft Preparation Report by Investment Centre, Food and Agriculture Organization (FAO)/World Bank Co-operative Programme, Rome in the States of West Bengal, Orissa, and Andhra Pradesh, the three east coast States of India. The second mission identified these six brackish water sites for a detailed socio-economic study. The study is expected to provide detailed information and analysis for consideration of farm operation and mangement. The study has been undertaken in consultation with mission officials, FAO, rural sociologist, FAO and detailed review of literature done by us. For this purpose Dr. Binod C. Agrawal of Indian Space Research Organisation (ISRO), Ahmedabad, visited Rome from November 4-11, 1990. During this visit, several discussions were held regarding the objectives of the proposed study, research methodology to be followed, time table of the study, human power requirements, and other details. Accordingly, a research proposal for the study entitled "Brackish Water and Inland Fisheries Project : Socio-Economic Aspects for Study" was prepared and the same was revised after comments. A copy of the draft proposal is enclosed in Appendix-1.

The research design and the actual methods and techiques followed in the study have been discussed in Chapter-2 of the report. The aim of this report is to provide asnwers to the following questions for taking decisions related to semi-intensive brackish water shrimp farming proposed in three States of India to be supported by the World Bank :

(i) How to involve the potentially poor individual beneficiaries in the brackish water semi-intensive shrimp farm operations and management to provide maximum benefits to them ?

(ii) What size of ponds should be designed for a given brackish water shrimp farm in each of the three states ?

(iii) What kind of techno-managerial support will be required to run the brackish water shrimp farm ?

(iv) Who should be responsible for masterminding the total operation, input supply, production, supervision and marketing of shrimp in each state ?

(v) How the benefits/losses will be shared by the individual, community and co-operatives, or even states ?

Keeping the objectives of the study in view, the time available and the six brackish water sites to be covered for collecting relevant socio-economic information, it was thought most appropriate to use Rapid Rural Appraisal (RRA) method. It was thought that by adopting this method we will be able to learn about the coastal rural conditions "in an intensive, iterative, and expeditious manner." (Community Forestry Rapid Appraisal 1989 : i).

In our own studies, we have been following a Quick and Useful Evaluative Research Investigation (QUERI) technique for more than a decade. The QUERI technique has been used in formative researches related to development communication. In television programme production meant for rural development, formative researches are specially designed to improve the television programme content, comprehension, and quality, that too in a limited time. Such formative researches aimed at reducing the uncertainties about effectiveness of the television programmes and thereby helped imrpove the quality of the programmes. Among many rapid appraisal techniques our team was familiar with QUERI technique. Hence, we combined rapid appraisal technique with our QUERI technique for the present research study.

In the present context, we wanted to understand the socio-economic condition of the potential beneficiaries in the selected proposed brackish water fisheries sites, their views and opinions about the pond size and methods of management, among other things. We needed focused and detailed information in this regard. The only thing known to us were the names of six brackish water sites selected by FAO mission. Except one change in case of Orissa, other five sites were identified in our proposal and were suggested.

The Report

The report has seven chapters. In Chapter 1, the background of the study has been discussed. In Chapter 2, the research methods and techniques followed in the study have been described. In Chapter 3, the integrated findings of the study have been presented. In Chapter 4, the suggestions and recommendations emerging from the integrated findings are presented. In Chapters 5, 6 and 7, analyses of individual sites of three states have been presented separately. In Chapter 5, the analyses of Dadanpatra and Digha sites of West Bengal have been presented. In Chapter 6, the analyses of Jagatjore and Panaspada sites of Orissa have been presented. In Chapter 7, the analyses of P.T. Palem and C.G. Palem sites of Andhra Pradesh have been presented. This sequence of presentation is based on the assumption that the reader can have a total picture after reading the first four chapters of the report. The remaining chapters can be referred to have details of individual sites. Six Appendices are meant to provide other details related to the locales of the study. The summary of the study has been provided in the begining of the report.

2

RESEARCH METHODS AND TECHNIQUES

In the resent study, we wanted to understand the socio-economic condition of the potential beneficiaries in the selected proposed brackish water shrimp sites, their views and opinions about the pond size and methods of management, among other things. We needed focused and detailed information in this regard. The only thing known to us was the names of six brackish water sites from among 14 proposed sites selected by FAO mission (see Map 1). Except one change in case of Orissa, other five sites were indentified in out proposal and were suggested for first hand investigation (see Appendix-1). Karanjamal brackish water site was replaced by Jagatjore brackish water site in the State of Orissa. Hence, the following brackish water sites have been surveyed :

	Name of the Brackish Water Sites	Name of the States
1.	Dadanpatra	West Bengal
2.	Digha	West Bengal
3.	Panasapada	Orissa
4.	Jagatjore	Orissa
5.	P. T. Palem	Andhra Pradesh
6.	C. G. Palem	Andhra Pradesh

For this survey, an intensive exploratory study design was used. Two stage purposive sampling was followed. In the first stage, the villages were selected. In the second stage, individual respondents were selected.

Village Selection

The village selection for this study was guided on the assumptions that : (a) beneficiaries can walk up to the maximum distance of five kilometres (km) to work on the brackish water fisheries site; (b) within five km radius there will be enough number of beneficiaries to participate in the project; and (c) the group and individual opinions and veiws will be representative of the site areas. Further, it was decided that the total number of villages covered for the survey should have at least three times more households as compared to the brackish water site size in a hectare (ha.). *For example*, if the proposed site is of 100 ha. the villages covered in the survey should have minimum of 300 households.

As a first step, District Census Handbooks of six sites were collected. From each census handbook, map of the smallest administrative unit in which the site was located was prepared with the held of the thematic maps. The smallest administrative unit is referred to as *Taluk* and *Tehsil*. In case of Andhra Pradesh, it is referred to as Taluk. In case of West Bengal and Orissa, the same has been referred to as Tehsil. In most cases Tehsil is equal to Police Station jurisdiction. All maps of Taluk Police Station of the six sites were obtained (see Maps 5.2, 5.5, 6.2, 6.5, 7.2, 7.5). In the census, Taluk/Police Station Maps, with the help of thematic maps, actual brackish water sites were located. Using the brackish water site as a centre, villages falling within the circle of five kilometres were identified. Though, in theory, they were in the radius of five kilometres, actual physical distances were more than five kilometres in some cases. In theory, it seemed to be a very sound decision for the survey. However, upon arrival at the sites, we observed the following :

In Andhra Pradesh, both P.T. Palem and C.G. Palem have several settlements which are part of revenue village comprising a total number of 15 settlements, eight in P.T. Palem and seven in C.G. Palem, officially referred to as hamlets. It would be essential to define at this stage, the concept of village in Indian terms. "A village refers to any tract of land recognized as a village by the local administration in the revenue record. The revenue village may have a definite surveyed boundary or not. The revenue village may comprise several hamlets which do not have definite boundary or areas distinctly marked. The revenue village is inclusive of the sites of residence of the inhabitants and the entire area including hills, *nalas*, cultivable land and village forest, but excluding reserved forest. An uninhabited village is one which does not have any residential population in the entire tract. A revenue village may comprise several hamlets with a boundary in case of a surveyed village or a notional boundary in case of unsurveyed village; but the entire village is one unit for presentation of census data." (District Census Handbook, Puri District 1981:1).

In case of West Bengal, revenue village is referred to as *Mauza* and *Tola* is referred to as hamlet. Same holds true in case of Orissa. In this study, all the revenue villages including hamlets were included for the purpose of socio-economic appraisal. It may be mentioned that some revenue villages are so large that they have several hamlets bigger than an average village of the area. Hence, they were surveyed as village. In this way, on an average, nine villages were surveyed in each brackish water site. The term village here means a rural settlement with defined special boundary having a distinct name for the purpose of reference and recorded in the official documents as such. Table 2.1 provides the distribution of villages/hamlets surveyed for the present purpose.

Table 2.1 : Distribution of Villages/Hamlets Surveyed by Brackish Water Sites

Sr. No.	*Name of Brackish Water Site*	*No. of Revenue Villages*	*No. of Hamlets*	*Total*
1.	Dadanpatra	11	0	11
2.	Digha	7	0	7
3.	Panaspada	10	0	10
4.	Jagatjore	12	1	13
5.	P.T. Palem	0	7	7
6.	C. G. Palem	1	6	7
	Total	41 (74.6)	14 (25.4)	55 (100.0)

Instrument Development

Based on the initial discussions in Rome with experts, the following schedule guides were prepared for data collection :

(1) Performa 1 : Village level data on caste, occupation and economy proforma
(2) Village schedule guide
(3) Case study of the organization of fishing groups within a village
(4) Group level interview schedule guide
(5) Household level interview schedule guide : Respondent-head of the household

The instruments prepared in Rome were revised after discussion with the research team in Ahmedabad. Proforma 1 was incorporated in the village schedule guide. The research team already familiar with the Quick and Useful Evaluative Research Investigation (QUERI) technique decided to follow the same in combination with rapid rural appraisal technique of data collection and evolved the following interview schedule guides :

Queri	1	Village schedule guide
Queri	2	Case study of the organization of fishing groups within a village
Queri	3	Group level interview schedule guide
Queri	4	Household level interview schedule guide

Language of the Schedule Guides

Given the linguistic diversity and data to be collected from the remote coastal villages, it was felt essential that data must be collected in the local langauge and in dialects spoken in the area. Also, the data must be collected uniformly for comparison. Hence, the research team decided to translate the schedule guides into the respective languages of the state (Bengali for West Bengal, Oriya for Orissa and Telugu for Andhra Pradesh). In Orissa, before translations were done the research team alongwith three project leaders and five Oriya speaking field investigators, discussed all the schedules guides item by item. In a way, the schedule guides were "pre-tested" in its first translation. Subsequently, the two project leaders got the schedule guides translated into Bengali and Telugu before starting the field survey.

Planning for Survey and Field Training

A research team of three Senior Social Scientists from ISRO were involved in the field study. At the first level these three members, namely, Binod C. Agrawal, Sham Behari Sharma and Arbind Sinha alongwith other Social Scientists of Development and Educational Communication Unit (DECU)/ISRO discussed the QUERI technique of data collection in detail at Ahmedabad while schedule guides were finalized. This detailed discussion helped the team to understand the aim and scope of the study. The team members also identified the important aspects of the study. During two weeks of the preparation period the team were involved in : (a) collecting secondary data, and thematic maps of the six brackish water sites, (b) detailed discussion with the Remote Sensing Team working on the project, (c) establishing contact with the Department of Agriculture, Government of India; Department of Fisheries of respective states and other officials connected with the project.

On Novemeber 22, 1990, Binod C. Agrawal left for Hyderabad to plan for field survey in Andhra Pradesh, to collect secondary data from the Department of Fisheries and to interview state officials. On November 26 and 27, 1990, the other two project leaders went to Delhi for contacting Government of India officials. On November 28, 1990, all three project leaders met in Cuttack, where five field investigators were selected with the assistance of Audience Research Unit, Television Authority of India who maintain a list of field investigators for field data collection. Two days were spent in translation of the schedule guides, planning of the field survey, training of the investigtors. In consultation with Brackish Water Fisheries Development Agency (BFDA), Cuttack, logistic arrangements like transport and accomodation were made on the following day. The project leaders tried to visit Mahakalpada Block to finalize the list of the villages within five km radius. However, the team still could not reach to Jagatjore brackish water site as waterway launch service was not available and was not feasible for the survey team to travel through that route. The team took other route by road and the actual sruvey started on December 1, 1990 only. From December 1, 1990, the field survey started first in Orissa, in which two project leaders fully participated as Binod C. Agrawal returned to Ahmedabad on December 2, 1990. After the survey in Orissa was over on December 10, 1990, one project leader moved to West Bengal and the others went to Andhra Pradesh. The same sequence of team recruitment and training was followed. In these two states, Department of Fisheries provided assistance in recruiting field investigators while logistic arrangements were made by the project leaders. All data collection was over by December 22, 1990.

The research team felt that even after detailed discussions and training, the field team though were effective in verbal communication, were not able to carry out an indepth probing of the respondents to a desirable level. Secondly, investigators tended to collect data treating the schedule guide as questionnaire. Given the nature of survey, in spite of our best efforts not a single woman investigator could be hired for this purpose, and the entire data were collected by all men team. In the coastal villages, the women were uninhibited while talking to male investigator. Hence, quality of the data did not get affected and women could be interviewed.

Data Collection

The data collection was divided into three parts. In the first part, all the officials in the Department of Fisheries were to be interviewed including the field level functionaries. In the second part, "village group interviews" were to be conducted. In the third part, selected individuals were to be interviewed.

Interviews of the Officials

The three project leaders were given direct responsibility to interview Department of Fisheries officials. The field investigators were asked to report about their own discussion/informal talks with the officials to the project leaders in the evening meetings held after the day's data collection. But during these meetings they were not able to exchange much views. However, during breakfast, lunch, dinner and while travelling, great deal of information were exchanged among the research team. The project leaders were also responsible for data collection from official sources like Block offices of the respective *Taluk*, Department of Fisheries and other development agencies. Also, they collected village level data with the help of field investigators and other Department of Fisheries / Block officials.

Village Level Group Discussion

Typically, upon arrival in a village the Fisheries Extension Officer, or the Fisheries Inspectors took responsibility of collecting adult men and women alongwith influential persons of the village. The group size varied from 15 to 30 persons in which only few women participated, that too, more like spectators than active participants. The smallest gatherings were observed in villages of Andhra Pradesh numbering 10 to 15 persons. With the help of the field investigators, the project leaders introduced the purpose of their visit to the group of villagers gathered. It was indicated that World Bank proposed to support brackish water fisheries in the vicinity. Before launching the project, it was thought essential to gather the opinions and views of the potential beneficiaries for its management and organization. The group was also told that it was proposed that no subsidy would be provided for this project by the World

Bank. Special mention was made about the size of individual fish ponds to be constructed for this purpose.

In most cases, it was observed that these group discussions were dominated by a few elected leaders or influential persons of the village. In spite cf our request, fisheries officials also influenced the thrust of the group discussion as discussed in Andhra Pradesh and on Jagatjore site of Orissa. The main points emerging out of the group discussion were noted by the project leaders in QUERI 3.

It may be mentioned that during the group discussions women generally did not speak even when they were present in the group. It was pointed out by several men that fishing is a man's job hence women will not express opinion in this regard. The village census data were collected with the help of village schedule guide (QUERI 1) prior to group discussion. However, in West Bengal, the same was done after the group discussion. The reason was convenience of the project leader. The entire village level data have been computed for future analysis and provided in Appendix-5.

The village level group discussions were initially organized with the help of Department of Fisheries officials, local elected leaders, and teachers. While official participation faciliated village level group discussions, their very presence to some extent gave the impression to the villagers that we were also "officials" and accordingly, the initial responses were "desirable" responses. Later project leaders and field investigators initiated group discussions.

Individual Level Data Collection

After the group discussion and village level data collection, the individuals were interviewed in their respective homes. The selection of the individual respondents were done on the purposive basis. All efforts were made to have representation of fishing and non-fishing heads of the households in the sample. Also that there should be women's representation in the sample. In spite of our best efforts, it was not possible to get an ideal sample as desired due to several factors including adults not being available at the time of interviews, and in few cases lack of approach to the women respondents.

Table 2.2 provides the final sample of respondents interviewed for this study. Table 2.2 shows that in West Bengal we could not get good response from women respondents. But on the whole we were able to get 58.2 per cent male respondents (see Table 2.2). In spite of all difficulties, male/female ratio was about 6:4 which was fairly good given the initial problems of having no female investigators in the team.

Table 2.2 : Distribution of Respondents by Sex and Brackish Water Site

Sr. No.	*Name of Brackish Water Site*	*Male*	*Female*	*Total*
1.	Dadanpatra	45	18	63
2.	Digha	27	13	40
3.	Panaspada	35	25	60
4.	Jagatjore	40	34	74
5.	P. T. Palem	18	23	41
6.	C. G. Palem	20	20	40
	Total	185	133	318
	Percentage	58.2	41.8	100

Initially, it was decided that individual respondents will be selected during group discussion. However, the same was not possible as group discussion could not attract women and on some occasions, a small number attended group discussion. Keeping in view the needs of the suggested kinds of respondents, the research team went in the village and directly contacted some individuals for interview. In such situations, the field investigators had to repeat the purpose of the survey to each respondent. It was observed that in many cases, the responses of the respondents were similar to those during group discussion. The research team on several occasions felt that individual replies were conditioned by the earlier group discussion in which the respondent was present as "spectator" or "listener". All the interviews were conducted in the language spoken by the respondents.

Case Studies

Using QUERI-2 a number of intensive case studies about the fisheries co-operatives were conducted. The data for these case studies were collected from the co-operative members and elected office bearers, Department of Fisheries co-operatives officials and from experts in the field of co-operatives. For this purpose, secondary data were also collected from the National Institute of Rural Development, Hyderabad.

Analysis

While most of the first level analyses were attempted in the field, the final analysis was done at Ahmedabad. In order to reduce error in the quantitative data, computer was used to the extent possible. Given the nature of data collected, no attempt was made to check any statistical validity. In most cases, the attempt was made of cross tabulate individual responses.

3

PROFILE, RESPONSES AND RESULTS FROM BRACKISH WATER SITES

In this Chapter, an attempt has been made to describe the profile of six brackish water sites. The profile of six brackish water sites includes the descriptive analysis of the sites, their current utilization and human habitation. Also, the socio-economic profile of the potential beneficiaries is described.

Three sets of data were available to locate, describe and analyse the profile of six brackish water fisheries sites, namely Dadanpatra (West Bengal), Digha (West Bengal), Jagatjore (Orissa), Panaspada (Orissa), P.T. Palem (Andhra Pradesh) and C.G. Palem (Andhra Pradesh) (see Map 1). First, actual site visits were made by our Remote Sensing and Social Sciences Research Teams based on thematic maps. Second set of data come from the official sources of the Department of Fisheries of the respective States. And, the third set of data were obtained from published reports like Census of India and District Census Handbooks. In addition, published and unpublished reports related to the present topic were also reviewed. Based on all data, an attempt has been made to answer some of the basic questions raised earlier.

The Sites

The individual sites as described in the Chapters 4, 5 and 6 clearly indicate that all the sites are away from any human habitation. Typically, the sites are flat mud surfaces and are used for either fishing related activities or salt making. In West Bengal, the two sites are being used for fishing. Both sites in Orissa are being used for prawn cultivation. In Andhra Pradesh, C.G. Palem site is being used unsuccessfully for salt farming. In P.T. Palem while, local fishermen are involved in prawn farming, the site is not being used for any other purpose. All the six sites are not directly approachable by any road nor through any public transport. The nearest site to the road side is that of Panaspada which is located within a km from a metalled road. In case of Dadanpatra and

Digha, approximately three kilometres distance must be covered to reach the metalled road. The most difficult site is Jagatjore where one cannot reach so easily. One has to cover more than 12-15 km by walk, cycle or skilled motor cycle ride to reach the site from a road. The two sites in Andhra Pradesh are within two km from the road side. But, these roads are seasonal roads and are available only during summer. However, they are about 12 km away from all weather roads.

Legal and Political Aspects of the Sites

While there is no doubt that each State Government can make the sites available for semi-intensive shrimp farming, it may not be out of place to mention the present legal and political situation of the site. While both proposed sites in West Bengal are free from any unauthorized occupation of use, some legal problems have been reported in the adjoining land of Dadanpatra site. Jagatjore site in under the control of Orissa Marine and Chilka Development Agency (OMCADA) who may not like to loose its right too easily. In case of Andhra Pradesh, while C.G. Palem site is under controversy as the villagers are claiming to use the site for salt farming, P.T. Palem in Andhra Pradesh and Panaspada in Orissa have no such problem. Some amount of political activities and involvement seemed to exist in most of the sites.

Human Habitation

The human population lives away form the proposed sites, the minimal distance is less than a km. In all sites, the potential human population who could directly benefit from semi-intensive shrimp farming are scattered in not less than a radius of five km. Observations indicate that fishing households in West Bengal live in *Khatti* or temporary settlements for a part of the year for fishing. Similar observations were made in Orissa and Andhra Pradesh. While there might not be a special provision for near the site housing of the individual fish farmers, some temporary shelters will be needed which can be constructed by the individuals on their own. What is needed is a provision of some open space near the ponds or sites for this purpose.

Socio-Economic Profile of the Coastal Population

Two sets of data, namely village census and individual responses have been utilized to describe the socio-economic profile of the villagers in the coastal villages covered under study.

The coastal population in three states, by and large, are closely associated with water — whether sea water or fresh water. Their total life centres around the river, and sea in one form or the other. The socialization process of the children reflects this close connection with water both as a source of livelihood and disaster that water brings in their lives. In an early age, children of both sexes begin to interact intensely with water, learn swimming, fishing, clearing of ponds and waterlogged land. It is quite common to observe children fishing and boating. Most of the learning is through imitative behaviour and participation in the activities with the elders. Regardless of caste and the caste occupation, an individual participates in fishing activities in a variety of ways. It is for this reason that even those who claim to be cultivators also practice fishing as shown in Appendix-5.

With the exception of a few households, the entire population living near brackish water sites are Hindus. For the purposes of analysis, this population has been divided into three "caste category groups" namely, "fishermen caste," "Scheduled Castes/Scheduled Tribes," and other castes. The first-two caste categories belong to the bottom of social hierarchy and are generally, economically poorest in the village, whereas, the category of other castes contains those who are at the top of the social hierarchy and are supposed to be economically sound and also own land and property. The caste distribution as indicated in Table 3.1 may not be a good representative of the caste distribution of the coastal areas, but it does reflect that majority are fisherman and people belonging to Scheduled Castes/Scheduled Tribes are involved in fishing (Graph 3.1). Both survey and village level data support this view (see Table 3.2 and Appendix-5). It may be mentioned that the category "Other" in case of occupation includes village artisans who also take active part in fishing alongwith cultivators and fishermen. It is for this reason that in the village census, multiple occupations have been indicated (Graph 3.2). In this respect the coastal population living near brackish water sites can be characterized as "fisher folk" whose life primarily centres around fish and fish related activities.

Several calculations were made to find out the ratio of cultivable land and human beings depending upon them. Based on the District Census Handbook 1981, an estimate was made for all the 55 villages under study. The man-land ratio did not exceed even one ha. while the average was below 0.5 ha. The ratio has been lowest for West Bengal followed by Orissa and Andhra Pradesh. Table 3.3 shows that 71.7 per cent respondents

Table 3.1 : Percentage Distribution of Respondents by Caste

Caste Category	*Sites*						
	Dadan-patra	*Digha*	*Jagatjore*	*Panas-pada*	*PT Palem*	*CG Palem*	*Total*
Fishermen	26	18	11	9	17	16	97
	26.80	18.56	11.34	9.28	17.53	16.49	100.00
	41.27	45.00	14.86	15.00	41.46	40.00	30.50
Scheduled Castes/ Scheduled Tribes	10	6	14	27	4	7	68
	14.71	8.82	20.59	39.71	5.88	10.29	100.00
	15.87	15.00	18.92	45.00	9.76	17.50	21.38
Other Castes	27	16	49	24	20	17	153
	17.65	10.46	32.03	15.69	13.07	11.11	100.00
	42.86	40.00	66.22	40.00	48.78	42.50	48.11
Total	63	40	74	60	41	40	318
	19.81	12.58	23.27	18.87	12.89	12.58	100.00
	100.00	100.00	100.00	100.00	100.00	100.00	100.00

Table 3.2 : Percentage Distribution of Respondents by Occupation

Occupation	*Sites*						
	Dadan-patra	*Digha*	*Jagatjore*	*Panas-pada*	*PT Palem*	*CG Palem*	*Total*
Fishing	28	25	12	35	17	12	129
	21.71	19.38	9.30	27.13	13.18	9.30	100.00
	44.44	62.50	16.22	58.33	41.46	30.00	40.57
Cultivation	20	2	39	9	14	15	99
	20.20	2.02	39.39	9.09	14.14	15.15	100.00
	31.75	5.00	52.70	15.00	34.15	37.50	31.13
Others	15	13	23	16	10	13	90
	16.67	14.44	25.56	17.78	11.11	14.44	100.00
	23.81	32.50	31.08	26.67	24.39	32.50	28.30
Total	63	40	74	60	41	40	318
	19.81	12.58	23.27	18.87	12.89	12.58	100.00
	100.00	100.00	100.00	100.00	100.00	100.00	100.00

reported having no land for cultivation. Only 28.3 per cent had land below one to three acres. Majority (57.5 per cent) own no domestic animals (see Table 3.4) and 69.5 per cent live in thatched huts (see Table 3.5). As an estimated annual household income below Rs. 5,000 cannot provide two square meals a day to its members then 82.1 per cent

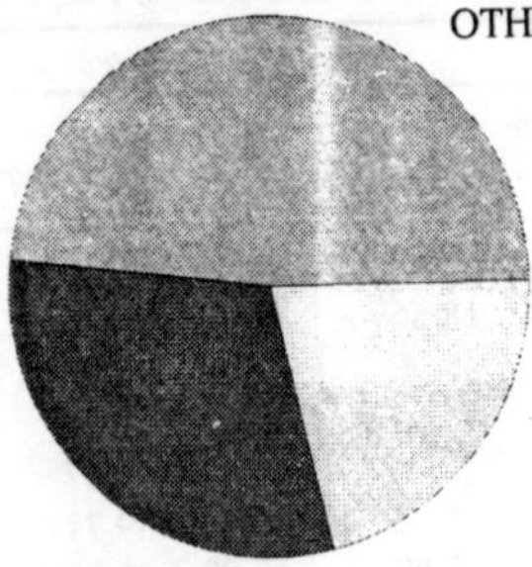

Graph 3.1 : Percentage Distribution of Respondents by Caste

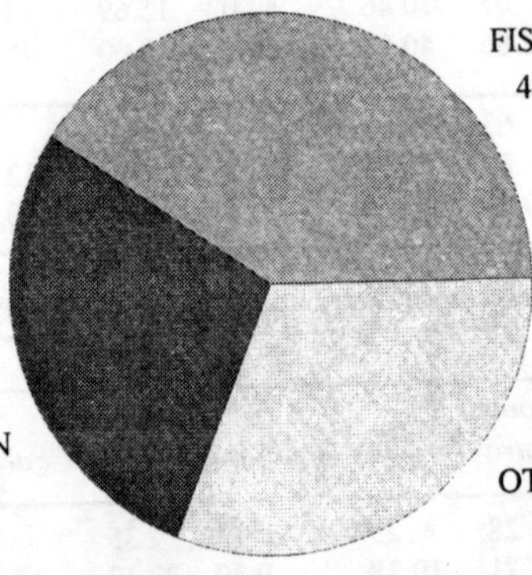

Graph 3.2 : Percentage Distribution of Respondents by Occupation

respondents should be considered economically poor or very poor (see Table 3.6). Given their economic condition, it is no surprise that 79.6 per cent have not taken loan or were not given loan by any one (see Table 3.7).

Table 3.8 shows that 46.9 per cent respondents are illiterate, there by indicating that 53.1 per cent respondents are literate. This figure is high when compared to the educational levels of all the three states. An analysis of the literacy rates of the *taluks* from where these respondents were drawn. Indicate somewhat lower literacy rate. The high rate can be attributed to two factors. One sample contained more men than women respondents : (58.2 per cent males as compared to 41.8 per cent females ; see Table 2.2). Two, the purposive sampling was in favour of educated persons who could express their opinion clearly. Table 3.9 provides the analysis of the mass-media exposure. The mass media exposure included radio, television, cinema and news papers.

Table 3.3 : Percentage Distribution of Respondents by Land Ownership

	Sites						
Land Ownership	*Dadan-patra*	*Digha*	*Jagatjore*	*Panas-pada*	*PT Palem*	*CG Palem*	*Total*
No Land	59	29	72	42	7	19	228
	25.88	12.72	31.58	18.42	3.07	8.33	100.00
	93.65	72.50	97.30	70.00	17.07	47.50	71.70
Below 1 Acre	1	9	—	8	4	—	22
	4.55	40.91		36.36	18.18		100.00
	1.59	22.50		13.33	9.76		6.92
1-3 Acre	2	2	1	10	25	18	58
	3.45	3.45	1.72	17.24	43.10	31.03	100.00
	3.17	5.00	1.35	16.67	60.98	45.00	18.24
Above 3 Acre	1	—	1	—	5	3	10
	10.00		10.00		50.00	30.00	100.00
	1.59		1.35		12.20	7.50	3.14
Total	63	40	74	60	41	40	318
	19.81	12.58	23.27	18.87	12.89	12.58	100.00
	100.00	100.00	100.00	100.00	100.00	100.00	100.00

Table 3.4 : Percentage Distribution of Respondents by Cattle Ownership

	Sites						
Cattle Ownership	*Dadan-patra*	*Digha*	*Jagatjore*	*Panas-pada*	*PT Palem*	*CG Palem*	*Total*
Cattle/Buffalo	14	8	46	11	22	17	118
	11.86	6.78	38.98	9.32	18.64	14.41	100.00
	22.22	20.00	62.16	18.33	53.66	42.50	37.11
No Domestic Animals	44	31	22	45	19	22	183
	24.04	16.94	12.02	24.59	10.38	12.02	100.00
	69.84	77.50	29.73	75.00	46.34	55.00	57.55
Any Other Domestic Animal	5	1	20	4	2	3	35
	14.29	2.86	57.14	11.43	5.71	8.57	100.00
	7.94	2.50	27.03	6.67	4.88	7.50	11.01
Total	63	40	74	60	41	40	318
	19.81	12.58	23.27	18.87	12.89	12.58	100.00
	100.00	100.00	100.00	100.00	100.00	100.00	100.00

Table 3.5 : Percentage Distribution of Respondents by House-Type

	Sites						
House-Type	*Dadan-patra*	*Digha*	*Jagatjore*	*Panas-pada*	*PT Palem*	*CG Palem*	*Total*
Thatched Hut	36	26	56	54	25	24	221
	16.29	11.76	25.34	24.43	11.31	10.86	100.00
	57.14	65.00	75.68	90.00	60.98	60.00	69.50
Kuchha	27	13	15	5	3	9	72
	37.50	18.06	20.83	6.94	4.17	12.50	100.00
	42.86	32.50	20.27	8.33	7.32	22.50	22.64
Pucca-Thatched	—	1	3	1	13	7	25
		4.00	12.00	4.00	52.00	28.00	100.00
		2.50	4.05	1.67	31.71	17.50	7.86
Total	63	40	74	60	41	40	318
	19.81	12.58	23.27	18.87	12.89	12.58	100.00
	100.00	100.00	100.00	100.00	100.00	100.00	100.00

Table 3.6 : Percentage Distribution of Respondents by Annual Income

	Sites						
Annual Income (Rs.)	*Dadan-patra*	*Digha*	*Jagatjore*	*Panas-pada*	*PT Palem*	*CG Palem*	*Total*
Below 3000	28	15	24	22	15	8	112
	25.00	13.39	21.43	19.64	13.39	7.14	100.00
	44.44	37.50	32.43	36.67	36.59	20.00	35.22
3000-5000	25	19	27	29	20	29	149
	16.78	12.75	18.12	19.46	13.42	19.46	100.00
	39.68	47.50	36.49	48.33	48.78	72.50	46.86
Above 5000	10	6	23	9	6	3	57
	17.54	10.53	40.35	15.79	10.53	5.26	100.00
	15.87	15.00	31.08	15.00	14.63	7.50	17.92
Total	63	40	74	60	41	40	318
	19.81	12.58	23.27	18.87	12.89	12.58	100.00
	100.00	100.00	100.00	100.00	100.00	100.00	100.00

Table 3.7 : Percentage Distribution of Respondents by Loan Taken

Loan Taken (in Rs.)	*Sites*						
	Dadan-patra	*Digha*	*Jagatjore*	*Panas-pada*	*PT Palem*	*CG Palem*	*Total*
Nil	58	35	59	60	15	26	253
	22.92	13.83	23.32	23.72	5.93	10.28	100.00
	92.06	87.50	79.73	100.00	36.59	65.00	79.56
0-1000	1	—	6	—	—	—	7
	14.29		85.71				100.00
	1.59		8.11				2.20
1001-2000	1	1	3	—	11	3	19
	5.26	5.26	15.79		57.89	15.79	100.00
	1.59	2.50	4.05		26.83	7.50	5.97
2001-3000	1	—	3	—	3	—	7
	14.29		42.86		42.86		100.00
	1.59		4.05		7.32		2.20
Above 3000	2	4	3	—	12	1	22
	9.09	18.18	13.64		54.55	4.55	100.00
	3.17	10.00	4.05		29.27	2.50	6.92
Total	63	40	74	60	41	40	318
	19.81	12.58	23.27	18.87	12.89	12.58	100.00
	100.00	100.00	100.00	100.00	100.00	100.00	100.00

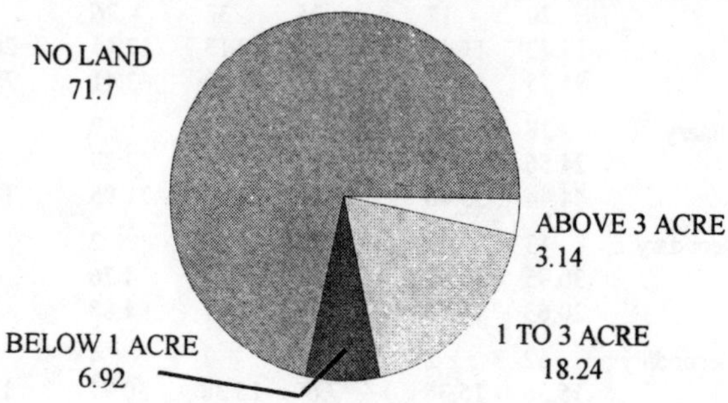

Graph 3.3 : Percentage Distribution of Respondents by Land Ownership

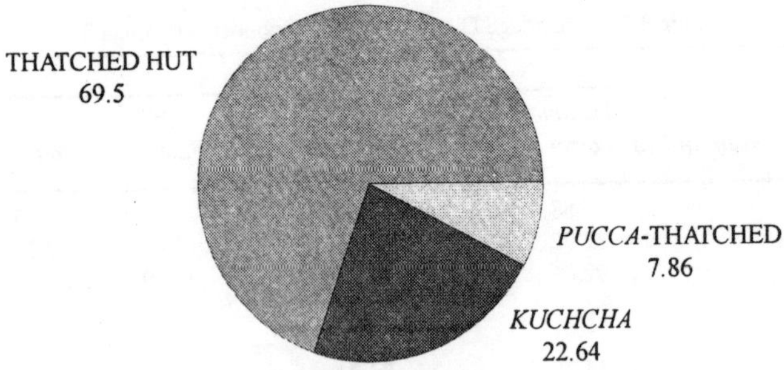

Graph 3.4 : Percentage Distribution of Respondents by House-Type

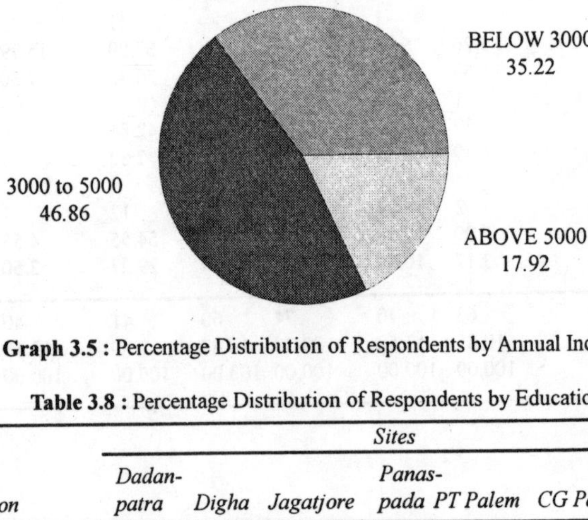

Graph 3.5 : Percentage Distribution of Respondents by Annual Income

Table 3.8 : Percentage Distribution of Respondents by Education

	Sites						
Education	*Dadan-patra*	*Digha*	*Jagatjore*	*Panas-pada*	*PT Palem*	*CG Palem*	*Total*
Illiterate	20	17	25	30	26	31	149
	13.42	11.41	16.78	20.13	17.45	20.81	100.00
	31.75	42.50	33.78	50.00	63.41	77.50	46.86
Below Primary	28	14	37	21	9	5	114
	24.56	12.28	32.46	18.42	7.89	4.39	100.00
	44.44	35.00	50.00	35.00	21.95	12.50	35.85
Below Secondary	13	7	11	7	2	2	42
	30.95	16.67	26.19	16.67	4.76	4.76	100.00
	20.63	17.50	14.86	11.67	4.88	5.00	13.21
Beyond Secondary	2	2	1	2	4	2	13
	15.38	15.38	7.69	15.38	30.77	15.38	100.00
	3.17	5.00	1.35	3.33	9.76	5.00	4.09
Total	63	40	74	60	41	40	318
	19.81	12.58	23.27	18.87	12.89	12.58	100.00
	100.00	100.00	100.00	100.00	100.00	100.00	100.00

Table 3.9 shows that 89 per cent respondents has either no mass-media exposure or poor mass-media exposure thereby indicating limited information about the outside world beyond the village and its vicinity.

Table 3.9 : Percentage Distribution of Respondents by Mass-media

	Sites						
Mass-Media	*Dadan-patra*	*Digha*	*Jagatjore*	*Panas-pada*	*PT Palem*	*CG Palem*	*Total*
Nil	19	13	46	42	9	8	137
	13.87	9.49	33.58	30.66	6.57	5.84	100.00
	30.16	32.50	62.16	70.00	21.95	20.00	43.08
Poor	34	16	17	17	32	30	146
	23.29	10.96	11.64	11.64	21.92	20.55	100.00
	53.97	40.00	22.97	28.33	78.05	75.00	45.91
High	10	11	11	1		2	35
	28.57	31.43	31.43	2.86		5.71	100.00
	15.87	27.50	14.86	1.67		5.00	11.01
Total	63	40	74	60	41	40	318
	19.81	12.58	23.27	18.87	12.89	12.58	100.00
	100.00	100.00	100.00	100.00	100.00	100.00	100.00

The socio-economic profile that emerges from the analysis is that the population living near six brackish water sites are "fisher folks" whose lives primarily centre around fish and fish related activities apart from limited amount of agriculture. Majority own no cultivable land, are illiterate and poor having household size between four to seven members with little mass-media exposure. The brackish water shrimp farming will be helpful in improving their economic condition.

4

SUGGESTIONS AND RECOMMENDATIONS

So far, we have discussed the details of the brackish water sites and the profile of the potential beneficiaries living within the five km radius. In this Chapter, as attempt will be made to answer the main questions raised in Chapter 1.

Semi-Intensive Shrimp Farming

The awareness and knowledge about shrimp and prawn farming, its economic benefits are known to almost all the villagers who participated in group discussion and who were individually interviewed regardless of their caste and current occupation. In this respect, acceptance of shrimp farming will be a welcome economic activity for the inhabitants of the coastal villages in all the three states. The 55 villages located within the five km radius of the brackish water fisheries sites cover an estimated population of more than 40,178 having a total number of 6,824 households. This shows an average household size of 5.9, though, it must be mentioned that the average household size in case of West Bengal is 7.4. Except in case of West Bengal, there will be no great difficulty in the selection of households wishing to participate in the semi-intensive shrimp farming. In case of West Bengal, only on Digha site, careful effort will be needed while selecting individual households for this project.

The total household socio-economic status analysis shows that more than 87.4 per cent households in these villages are "poor to the extent that they cannot earn enough to get two meals a day (seeTable 4.1). To a large extent, the same holds true in case of Middle Group also. This general picture is quite comparable with the sample survey data of the individual respondents drawn from the same villages (see Table 3.6). Exceptions apart almost all the households of these villages are poor. Hence, semi-intensive shrimp farming will provide them direct economic benefits. In this respect, semi-intensive shrimp, farming will achieve an important social goal of the country.

Table 4.1 : Distribution of Villages by Caste Category, Population, and Socio-Economic Status

Sr. No.	*Name of Site*	*Caste Category*	*Total Population*	*Total House-holds*	*Socio-Economic Status*		
					Lower	*Middle*	*Upper*
1.	Dadanpatra	Fishermen } SC/ST*	2230	280	280	0	0
		Other castes	8645	1157	1082	75	0
2.	Digha	Fishermen	300	25	25	0	0
		SC/ST*	1850	246	241	5	0
		Other castes	2375	383	376	12	0
3.	Panaspada	Fishermen	2075	299	129	150	20
		SC/ST*	1188	204	155	37	12
		Other castes	2909	438	281	154	3
4.	Jagatjore	Fishermen	560	96	90	6	0
		SC/ST*	2145	439	434	5	0
		Other castes	5190	879	666	190	23
5.	P.T.Palem	Fishermen	1350	316	316	0	0
		SC/ST*	654	155	148	5	2
		Other castes	2026	464	326	84	54
6.	C. G. Palem	Fishermen	4003	893	893	0	0
		SC/ST*	841	231	231	0	0
		Other castes	1837	319	292	27	0
	Total	Fishermen	8238	1629	1453	156	20
				100.00	89.20	9.58	1.23
			20.63	23.87	24.36	20.30	17.54
		SC/ST*	8908	1555	1489	52	14
				100.00	95.76	3.34	0.90
			22.17	22.79	24.96	6.93	12.28
		Other castes	22982	3640	3023	542	80
				100.00	83.05	14.89	2.20
			57.20	53.34	50.68	72.27	70.18
		Grand Total	40178	6824	5965	750	114
				100.00	87.41	10.99	1.67
			100.00	100.00	100.00	100.00	100.00

* SC/ST denotes Scheduled Castes/Scheduled Tribes.

Co-operatives and Co-operation for Semi-Intensive Shrimp Farming

There are differing perceptions about the question as to now semi-intensive shrimp farming should be done. Earlier four different approaches were suggested in the FAO mission report. Keeping these approaches in view indepth interviews of Fisheries Officials were conducted. In addition, the opinions of the villagers were sought to answer this question.

It must be mentioned that fisheries co-operative exist in the vicinity of each of the proposed sites. Out of 55 villages surveyed, 31 villages have fisheries co-operatives. Apart from these, there are a few unregistered co-operatives also. It must be mentioned that other kinds of co-operatives also exist in these villages (see Appendix-5).

Department of Fisheries of each State has separate Fisheries Co-operatives specially designed for fisheries activities. In oirder to smoothen the activities of these co-operatives, the District Fisheries Officer acts as Assistant Registrar of the co-operatives. For each kind of fishery activity like inland fisheries separate co-operatives have been formed. Some of these co-operatives are very active as indicated in the selected case studies on co-operative (see Appendix-6 and also Chapters 5, 6, and 7).

There are two kinds of views expressed by the State Government officials about semi-intensive shrimp farming. According to one view, this kind of high cost intensive and technically sophisticated farming should be undertaken by private companies of corporations. It is suggested that they will have appropriate management, manpower and resources for this purpose. In long-term, such an approach will be beneficial to the villagers also who will get employment with the company/corporation. In this way, the department of fisheries will have to deal with fewer people. We were told that without privatisation of shrimp farming, it will be extremely difficult to complete in the international market. Also, there was some sense of inter-state competition. At least in one state, the officials, indicated that another state was doing well in prawn farming because of the policy of letting private companies run prawn farming. Also, some of the officials have no faith in co-operatives. They mentioned about non-functioning co-operatives, indifference of the individual members towards co-operatives and rampant corruption in this institution. But a much larger group of officials strongly advocated for co-operative farming. Their argument was that in conformity with the existing government policy and national goals, it is very important to make all efforts to develop co-operatives in various sectors of the village to help poor.

According to second view expressed by the fisheries officials, supported in the group discussions and by individual respondents, the experiences of the co-operatives by and large have been good especially in fisheries. In this all members tend to benefit directly from the co-operatives. Semi-intensive shrimp farming as proposed will directly benefit individual households. It is, therefore, suggested that co-operatives for semi-intensive shrimp farming should be organized in all the three states on all the six sites. In each site, according to State Co-operative Laws, these co-operatives should be organized.

Individual and group opinions were sought regarding the membership of these co-operatives. In order of preference, it was suggested that the co-operatives should be organized by the members of the same caste. The second dominant view was that it should be organized in which the membership should be restricted to the individuals of the same village or villages within the Gram Panchayat. Should there be separate membership of men and women in the co-operative society? Observations indicated that even when fisheries ponds were allotted to women in Venkatadurgambapuram, Andhra Pradesh they were effectively run by men in all respects. Hence, it is suggested that a couple instead of either a man or a woman should be made member of the co-operative society. The rationable is that both men and women participate in shrimp farming, hence it should be owned by the couple. It will also give recognition to women and their important role in the semi-intensive shrimp farming.

Pond Size and Individual Ownership

One of the most important parameters having direct economic and technical implications and far reaching social ramifications has been the size of the individual ponds. At all levels, in every possible manner and situation, the question of the pond size was raised by the research team in all the three states. There has been no definite answer to the question except one fisheries official who argued that each pond should be of one ha. size. His argument was that a 0.5 ha. pond in reality is of 0.4 ha. size. This reduces the effective use of the pond and thereby reduces shrimp production. At the same time, this kind of design would cost more money. He suggested that a standard pond of one ha. should be constructed and if need arises two households should be asked to jointly farm on one ha. pond.

At present, all ponds constructed by the State Government in each state vary from 0.2 ha. to 0.4 ha. In no place has a one ha. pond been designed or constructed under any scheme of the Scheduled Castes Corporation (SCC), Integrated Rural Development Programme (IRDP) and

Economic Rehabilitation of Rural Poor (ERRP).

During group discussions and individual interviews one of the main questions asked by the research team was : Who will own the pond? It seems as if this has been upper most in the minds of the villagers. Such a question of ownership of pond or land is very legitimate and natural in the Indian context where ownership has great social and psychological meaning for an individual household. It may be important to note that the "Zamindari abolition" became the key issue in post independent India. All efforts have been made since then to give the land to the tiller. Ownership of land provides greatest social and psychological security to most of the Indian farmers. Hence, ownership of pond must be viewed in this perspective.

Therefore, the question of pond size should be examined in the light of two issues : (a) How maximum number of households can be given access to new economic opportunity, and (b) what should be the optimum size of the pond that an individual household can manage as a family enterprise or with the existing family human power. It is assumed that the household size is about six members including young children. The answer to the first question is rooted into the policy of the government which desires to have distributive justice and sharing of resources among the largest number of poor persons. "The *goal* of the project is to participate in the poverty alleviation thrust of the Government, benefitting the weaker sections and poorest of the poor in coastal zones marginal areas by introduction of semi-intensive shrimp where feasible" (FAO Report Working Paper : 4). To that extent, the design of the pond should be such that it can satisfy the largest number of poor households. The indirect but clear answer to the second question comes from the survey data. Table 4.1 (a) shows that 87.4 per cent respondents want to manage their ponds independently. Based on State specific experiences, about 15 or 4.7 per cent respondents living in P.T. Palem wanted to work on fixed wages and similarly 10 or 3.2 per cent Dadanpatra respondents wanted to work as co-operative members. If these two specific responses are separated, then more than 95 per cent respondents wish to own a pond of their own.

Table 4.2 shows that 67.0 per cent respondents feel confident that they can take care of their ponds independently whereas 33 per cent think that they would need assistance from others. However, all of them require assistance in the areas of finance, seed and feed and not dependency to become a part of a collective farming. The analysis of Table 4.2 further supports the view that there is a genuine desire on the part of respondents to have an independent pond owned by them for shrimp farming.

Table 4.1 (a) : Percentage Distribution of Respondents by Pond-Handling

	Sites						
Pond-Handling	*Dadan-patra*	*Digha*	*Jagatjore*	*Panas-pada*	*PT Palem*	*CG Palem*	*Total*
Independent	49	36	69	58	26	40	278
	17.63	12.95	24.82	20.86	9.35	14.39	100.00
	77.78	90.00	93.24	96.67	63.41	100.00	87.42
Fixed Wage	4	2	5	1	15	—	27
	14.81	7.41	18.52	3.70	55.56	—	100.00
	6.35	5.00	6.76	1.67	36.59	—	8.49
Work on Co-operative Basis	10	2	—	1	—	—	13
	76.92	15.38	—	7.69	—	—	100.00
	15.87	5.00	—	1.67	—	—	4.09
Total	63	40	74	60	41	40	318
19.81	12.58	23.27	18.87	12.89	12.58	100.00	
	100.00	100.00	100.00	100.00	100.00	100.00	100.00

Table 4.2 : Percentage Distribution of Respondents by Pond Care

	Sites						
Care of Pond	*Dadan-patra*	*Digha*	*Jagatjore*	*Panas-pada*	*PT Palem*	*CG Palem*	*Total*
Can Take Him-self/Herself	40	24	48	42	31	28	213
	18.78	11.27	22.54	19.72	14.55	13.15	100.00
	63.49	60.00	64.86	70.00	75.61	70.00	66.98
Assistance Required	23	16	26	18	10	12	105
	21.90	15.24	24.76	17.14	9.52	11.43	100.00
	36.51	40.00	35.14	30.00	24.39	30.00	33.02
Total	63	40	74	60	41	40	318
	19.81	12.58	23.27	18.87	12.89	12.58	100.00
	100.00	100.00	100.00	100.00	100.00	100.00	100.00

In the light of these answers and no expressed opinion about the pond size, it is easy to conclude that the pond size should be such which can be managed by the members of the household. Also majority of the respondents think that they can manage a 0.5 ha. pond. Based on the experiences in the site areas, it is, therefore, suggested that pond should be of 0.5 ha.

Selection of Criteria

One of the important questions raised in this study is : How to select a beneficiary for semi-intensive brackish water shrimp farming? Most of the data collected to answer this question were in the form of observations, interviews and quantitative estimations of the socio-economic conditions of the villagers living around and near the proposed brackish water sites alongwith a question raised in the group discussion and during individual interviews. There are several observations that are pertinent while discussing selection criteria of the beneficiaries.

First, more than 98 per cent of the village population surveyed have been characterized as poor. Second, a number of fisheries projects supported by Central Government, State Governments and UN Development Agencies under various poverty alleviation programmes are in operation. Third, the sites and some of the population falling within the proposed semi-intensive brackish water shrimp farming area by World Bank Project are beneficiaries of other on-going projects. In all these projects, subsidy of one kind or the other have been provided. Regardless of caste, religion and current occupation, a large majority of respondents were interested in semi-intensive brackish water shrimp farming in their individual ponds. The officials are of divergent views about the participation of the poor in this project. While some would like that such a project should be given to private companies to run, the others would like the poor to participate in it. What does it all amount to ? There is no simple single method of selecting beneficiaries for semi-intensive brackish water shrimp farming.

It is suggested that first preference should be given to those who have not been selected for any such development programme in the past. Second, in order to preference among three caste categories, the first preference should be given to fishermen caste followed by Scheduled Castes and Scheduled Tribes. Third, preference should be given to the economically poor other castes. As there are going to be some job openings at the farm levels, young educated persons within the site area should be employed and send for training.

So far, there is no fisheries development programme in the States of West Bengal, Orissa and Andhra Pradesh without a subsidy of one kind or the other provided by the government or sponsoring agency. Therefore, both for Department of Fisheries officials and participating fishermen and villagers, it was difficult to conceive of a situation in which a programme like semi-intensive shrimp farming can be initiated without any subsidy. In the past, subsidies have been provided in terms of full or part cost of the pond construction, purchase of boats, nets and nylon yarn. In some places, even the responsibility of construction of pond was taken directly by the government. *For example*, in P.T. Palem site area, ponds were constructed by the District Rural Development Agency from the funds provided under Integrated Rural Development Programme. Officials think that fisheries projects like shrimp farming cannot be implemented without subsidy. As a matter of fact, they do not raise this kind of question. To them such a project can only be considered with some financial and technical/extension support of the government.

After careful explanation of the brackish water shrimp farming project to the villagers during group discussion and even during individual interviews, an attempt was made to inform that there might not be any subsidy for this project from the government side. Not many questions were raised by the participants of the group discussion. However, they seemed keen in semi-intensive brackish water shrimp farming. They showed interset in knowing about the technical complexities of shrimp farm. After hearing and discussing all the details, the most obvious question raised by the villagers was about the cost invoved in the project. The next question asked related to the method of repayment of the loan or the cost of the construction of the individual ponds, and operating cost of the infrastructure facilities of the farm. Also, several questions were raised about seed and feed supply.

It seems no one was worried about the marketing aspects of the shrimp. Given the demand and high cost of the shrimp and the existing marketing network, it has been taken for granted that shrimps will sold at the door steps of the farmer or near the farm. Presently, commission agents, middle men and companies make arrangements for buying the fish, prawns and shrimps near the farm on a regular basis. The team observed that even selling a single prawn was acceptable to the buyers. Given the economic benefits from the shrimp farming, it is not difficult to except that semi-intensive shrimp farming will be profitable to the fishermen and non-fishermen even when the risks are high due to cyclone and other natural problems. Therefore, it was not at all surprising when the villagers asked about the methods of repayment instead of subsidy to be provided by the government.

Given the support for the construction of the ponds and development of the infrastructure facilities of the farm, direct subsidy may not be required in this project. However, given high risk requirements of several inputs like seed and feed, power supply and need for the management of the farm, it is recommended that these facilities should be partially supported. Insurance facilities against natural disaster, and crop should be provided as it has been observed that in the absence of such safegaurds, the individual fish farmers have suffered a great deal. Also the cost of technical support services, training and management should be provided with the support of the existing government institutions and organizations. This will provide indirect subsidy to the project.

Women's Participation

In the villages near brackish water sites, women play an important economic role in addition to their domestic roles. It was observed that women directly help in the preparation of the ponds, cleaning of the ponds, wild seed catch, fishing and similar fish related activities. In Jagatjore (Orissa), and in West Bengal women collect natural prawn seeds. It may be mentioned that tradionally Bengali women both in West Bengal and Orissa do this job as there is a small migrant Bengali population in the coastal villages of Orissa (see detail description in Chapter 6). Both men and women believe that women can play an important role in brackish water shrimp farming. Two main areas of women's participation identified by the respondents are feeding and guarding of the ponds alongwith the preparation of feed.

In multiple response to the question how women can help in shrimp farming, both men and women identified "farm guarding" as number one job (51.2 per cent) followed by feeding (46.5 per cent). Not more than 12.9 per cent thought no role for women in this venture (see Table 4.3). Further analysis shows that while women respondents see a role for themselves in brackish water shrimp farming, it is the men who see no important role for women. In the light of the suggestion made for providing independent small ponds of 0.5 ha. to individual households, it is necessary that women must play a vital role in the total management and operation of the individual ponds and farm. Also regular and sustained efforts are required for brackish water shrimp farming, so that the couple and not the man alone can take care of the pond. There is a need for division of labour within the household in which women can do a great deal to increase the productivity of the shrimp farm. Hence, it is suggested that the training programme must have a women's component focusing on the women's role in the shrimp farming.

Table 4.3 : Percentage Distribution of Respondents by Women Help

Women Help	*Dadan-patra*	*Digha*	*Jagatjore*	*Panas-pada*	*PT Palem*	*CG Palem*	*Total*
Feeding	41	28	37	25	39	25	195
	21.03	14.36	18.97	12.82	20.00	12.82	100.00
Farm Guarding	40	17	31	25	11	25	149
	26.85	11.41	20.81	16.78	7.38	16.78	100.00
Marketing	1	1	14	5	2	5	28
	3.57	3.57	50.00	17.86	7.14	17.86	100.00
House Wife	5	3	21	23		—	52
	9.62	5.77	40.38	44.23			100.00
No Reply	1	—	4	1	—	—	6
	16.67		66.67	16.67			100.00

'N' is more because multiple choice.

Organization and Management of the Brackish Water Shrimp Farm

From all indications, the psychological and social need to own or lease a pond was the uppermost in the mind of the villagers. As indicated earlier, the first choice of the respondents was to farm independently. Both these are cultural needs as well. Ownership of land and pond provides social status as well as a sense of security. It also provides a sense of fulfillment and means of livelihood. Hence, a management structure that gives a symbolic and concrete expression of these needs will be most acceptable to the potential beneficiaries. On the other hand, a centralized management at the farm level is essential to run the shrimp farms economically, efficiently with reasonable profits. Between these two requirements though, not necessarily opposing each other, a management system needs to be evolved. There are three main sub-systems in the hierarchy of management of a shrimp farm :

(a) Management of all aspects of a semi-intensive brackish water shrimp farm;
(b) Management of interface with external agencies for inputs, supplies and marketing; and
(c) Management of individual ponds.

Given the suggested size of the pond of half ha. it is clear that the farming responsibility of the ponds will be that of the individual households. The household members will be responsible for feeding, guarding, ensuring of the change of water, procurement of seed, and all related activities. Given the complex nature of the operation of the semi-intensive brackish water shrimp farming as compared to the current practice of prawn culture cultivation, it will be essential that the entire household members should be drawn into the shrimp farming and accordingly they should be trained.

In the light of the above discussion, it is suggested that individual households should be given farming rights of the pond or lease of ponds on long-term basis by the government. These individuals should then form the primary fisheries co-operatives as per rules of the state that are already in existance. If need arises, a few modifications should be incorporated in the existing fisheries co-operative rules. This primary fisheries co-operative will then with the help of experts and Department of Fisheries officials will create second level of management organization for day-to-day management. Initial capital required for this purpose will be part of the loan provided for this purpose. Same holds true in case of third level of management organization. At the highest level, the management of the farm should be handled by Board of Directors of the pattern of Indian Farmer's Fertilizers Co-operatives (IFFCO) Ltd. In this way, a farm takes the shape of an industrial unit managed by the professional team of experts. At the lowest level, the individual ponds will be treated as smaller production cell of the unit.

There is no doubt that there are divergent views on the management and organization of the semi-intensive brackish water shrimp farm. The horizon of individual fish farmers, potential beneficiary and fishermen is limited to their expectations from it in terms of getting pond rights for shrimp farming, load and subsides from government. The officials on the other hand think in terms of a scheme or project implementation, subsidy distribution, extension activities to create fish farming co-operatives. Given the importance attached to brackish water fisheries, separate agencies, cells and sections are created in the Department of Fisheries in the coastal states of West Bengal, Orissa and Andhra Pradesh.

Training Needs

At all levels, in every discussion, training needs were highlighted for the success of the project. While there were different kinds of motivations involved in suggested training, there is no denial that training must form, as suggested in the FAO mission report, an integral part of the project. Based on "felt needs" as expressed by the potential beneficiaries, and fisheries officials and observed needs, it is recommended that three types of trainings are required. These are :

(a) Technical and skill training both at farm and pond levels to all levels including the beneficiary couple (both men and women).
(b) Training in co-operation and co-operatives to all the beneficiary couples. This will help improve the functioning of the co-operatives, develop leadership qualities and role relationships between the farm management functionaries and individual shrimp farmers.
(c) Management training for individual farm staff and shrimp farmers for the total understanding of the organizational aspects of the shrimp farm apart from technical aspects of shrimp farming.
(d) Strengthening of extension services for the project to increase two way communication between the farmers and the Department of Fisheries and other external agencies.

5

WEST BENGAL

In this Chapter, two brackish water sites, namely Dadanpatra and Digha have been described. For each site separate socio-economic analyses has been presented. Efforts have been made to discuss major issues concerning the semi-intensive brackish water shrimp farming.

BRACKISH WATER SITE : DADANPATRA

Dadanpatra brackish water site is located in the revenue village of Daskhin Purushottampur of Ramnagar Police Station, District Medinipur. The Brackish Water Site Dadanpatra is near Contai — an important town. The villages surrounding the site (five km radius) form part of Police Station Ramnagar. (Map 5.1 & 5.2). The block headquaters is located on Digha-Contai road at a distance of 14 km from site village. The villages surrounding the site are approachable by network of Mooharam/*Kuchcha* road and water ways. The proposed site is located at village Dakshin Purushottampur, which is approachable through Saulakhal by Mooharam road and then one has to cross the sea dyke and walk depending upon high and low sea tide water and then cross the creek by country boat. While the country boat is easily available, walking on the land having sea weeds is quite difficult. The other site villages of Dadanpatra are approached through three different Mooharam and Kuchcha link roads through Digha-Contai road. Digha-Contai road is linked in the south with Balasore and in the north with Calcutta and towards west with Medinipur.

Profile of the Villages

The pattern of village settlement in Dadanpatra brackish water site is generally small and scattered and stretched along the road (see Map 5.3). Within a radius of five km from the brackish water site, there are a total of 11 revenue villages excluding 2 *khatti settlements* (seasonal camps). All the site villages except one falls in the jurisdiction of Ramnagar Police Station (PS). One-third villages of this Ramnagar PS are small (less than 500),

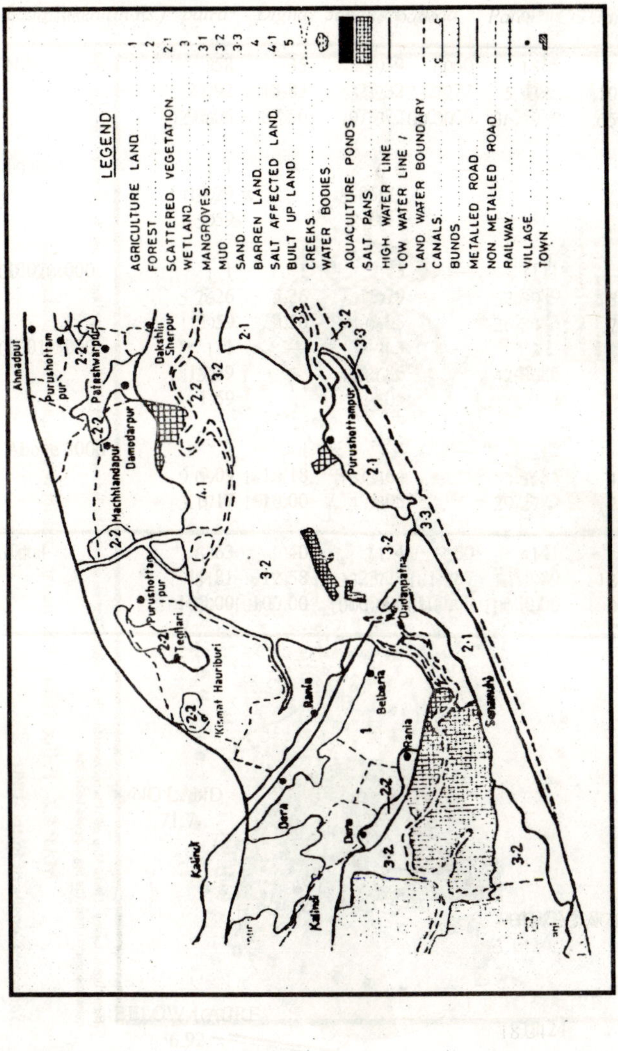

MAP 5.1 : Land use Map Site : Dadanpatra

Source : Remote Sensing Applications Group, ISRO, Ahmedabad, 1990.

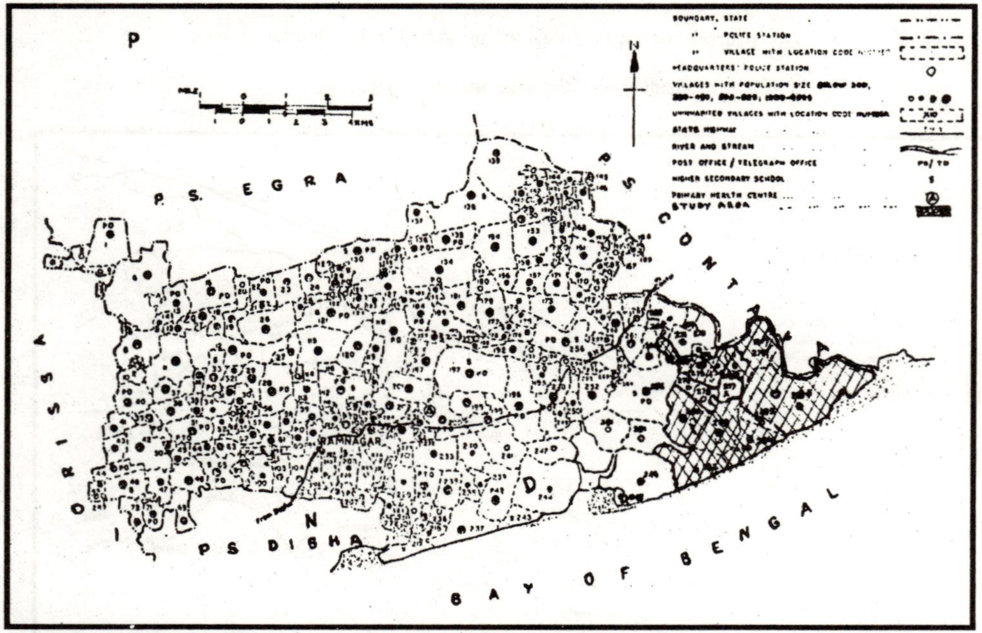

Based upon Survey of India Map with the permission of the Surveyor General of India. The Territorial States of India extended into the Sea to a distance of twelve nautical miles measured from the appropriate base line

MAP 5.2 : Police Station Ramnagar, District Medinipur Site : Dadanpatra

Source : Census of India, 1981.

while two-third villages are generally of a medium size. Two-third of the total geographical area of Ramnagar PS is available for cultivation and the percentage of irrigated area to total cultivable area is quite low (7 per cent). However, the situation with regard to irrigation of the site villages is slightly better as some of them have diesel pump sets for irrigation.

The site villages in terms of approach road are not so good. Most of them are at a distance of 5-6 km from Digha-Contai road, where public transport is available. Again Dadanpatra Brackish Water Site villages are connected by *kuchcha* mooharam road with the availability of manually driven cycle rickshaws. The availability of drinking water, electricity and educational facilities are better in the site villages. Private medical facilities are also available in some of the site villages. However, Contai town is also not very far off for such medical facilities. Agriculture is the predominant occupation in the site villages. Equally or even more important occupation is the fishing. Betal leaves, salt making, rickshaw pulling and other labour work in Contai/Digha Town are other important occupations. None of the site villages is inside the sea dyke except Dakshin Purushottampur. The availability of high lands with alluvial soil together with the sea dyke for its protection, availability of diesel pump sets for irrigation and nearness to sea coast as well as proximity to the town and frequently available public buses on Digha-Contai road, a hatchery and fisheries office at Digha goes to indicate that the site has good potential for a quick promotion of brackish water fish production.

Analysis of Individual Responses

In Dadanpatra site, sixty-three respondents spread over eleven villages were interviewed at the household level. Mostly the respondents were from those who had attended and participated in the group discussion. As regards female respondents, most of whom in spite of repeated persuasion either did not come or remained a mute listener at an unsually long distance during the group discussion. They had to be, therefore, contacted separately at individual level. Likewise sometimes even those make respodents had to be interviewed, who had not come for the group discussion. Table 5.1 indicates that 11 villages selected for the study comprised at total of 1437 households. 19.49 per cent of these households are of Fishermen including SC/ST castes and 80.51 per cent are from other caste groups. The average household size estimated through the triangulation technique comes out to 7.6 (on the face, this appears to be on higher side). Socio-Economic Status (SES) of 94.29 per cent of the households on lower side;

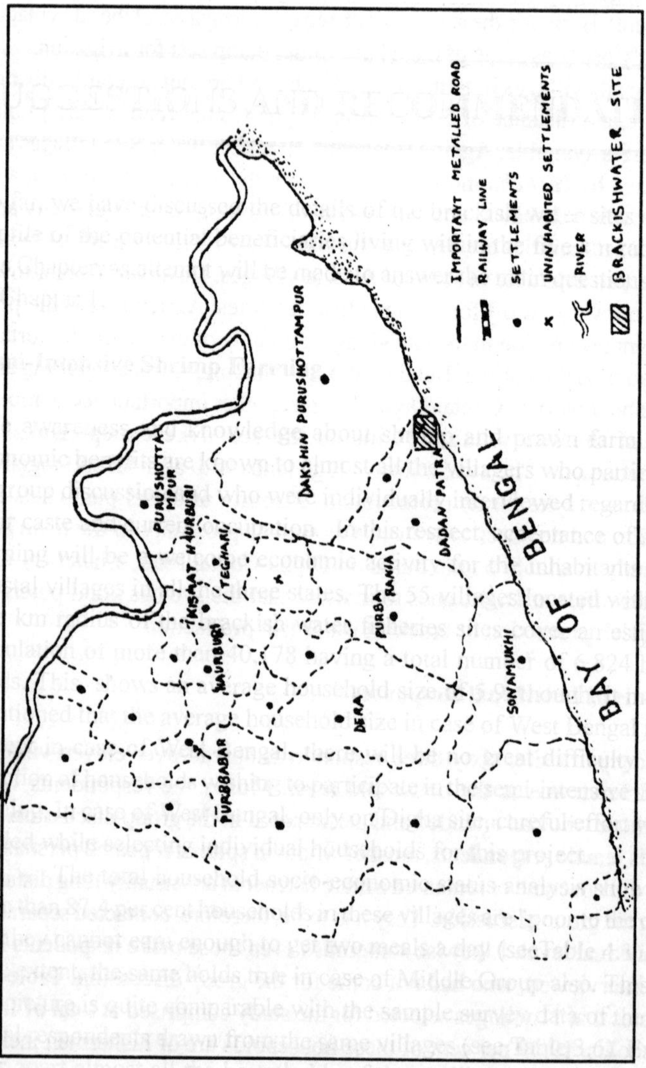

MAP 5.3 : Map showing villages visited at Dadanpatra Brackish Water Site

only very few households of middle and none of them is on upper side. There is no significant difference of SES among the fisherman including SC/ST and the other caste groups. With this background, an effort is now made to know the extent to which the respondents represent the targeted potential population of that area. However, before any inferences are drawn for the study, it is necessary to know the extent to which the sampled respondents represent the targeted potential population of the project.

Table 5.1 : Distribution of Caste Category by Households, Population and Socio-Economic Status in Study Villages

Site: Dadanpatra

Caste Category	*Number of Household*	*Total Population*	*Household Size*	*Socio-Economic Status* *Lower*	*Middle*	*Upper*
Fishermen/	280	2230	8.0	273	7	
Scheduled Castes/	100.00			97.50	2.50	
Scheduled Tribes	19.49	20.51		20.15		
Other Castes	1157	8645	7.5	1082	75	
	100.00			93.52	6.48	
	80.51	79.49		79.85	91.46	
Total	1437	10875	7.6	1355	82	
	100.00			94.29	5.71	
	100.00	100.00		100.00	100.00	

Table 2.2 indicates that almost two-third of the respondents are male. Table 5.1 indicates that 16 per cent of the respondents belong to Schedule Castes and 41 per cent are fishermen. Almost all of them are Hindus and only three from Muslim community. It further indicates that most of them have below secondary level of education, one-third of them are illiterate and rest are below primary. Little more than half of them are young *i.e.* up to 40 years (Table 5.11). Occupation-wise, the predominant occupation of half of the respondents is fishing; one-third agriculture and remaining others engaged in labour including agriculture labour (Table 5.3). Little less than two-third of the responents stay in thatched houses and the remaining little more than one-third have *kuchcha* houses with country made tiles on their roof (Table 5.4). Only few respondents own individual ponds and lands (Table 5.15 & 3.3), but one-fourth of them possess cattle (Table 3.4). Almost none of them have taken loan from any institution (Table 3.7). Income-wise, little less than half of the respondents earn average annual income below Rs.3,000 and little less than half of them earn up to Rs 5,000 and only a few of the respondents

earn more than Rs. 5,000 per annum (Table 5.2). Elsewhere it was observed that most of the respondents have very poor/poor exposure of mass-media. All the above together with analysis of overall village level data (Table 5.1) goes to indicate that the respondents are generally from poor population and are from varied group of Scheduled Caste, fishermen and other caste groups. This is because one of the important occupation of most of the other caste group people is fishing. The number of female respondents is only one-third as they have very limited role in deep sea fishing and its marketing at the fish landing sites.

Table 5.2 : Percentage Distribution of Respondents' Occupation by Annual Income

(Site: Dadanpatra)

Occupation	*Annual Income*			
	< 3000	*3000-5000*	*Above 5000*	*Total*
Fishing	16	11	1	28
	57.14	39.29	3.57	100.00
	57.14	44.00	10.00	44.44
Cultivation	5	10	5	20
	25.00	50.00	25.00	100.00
	17.86	40.00	50.00	31.75
Other Occupation	7	4	4	15
	46.67	26.67	26.67	100.00
	25.00	16.00	40.00	23.81
Total	28	25	10	63
	44.44	39.68	15.87	100.00
	100.00	100.00	100.00	100.00

Table 5.3 : Percentage Distribution of Respondents' Caste Category by Occupation

(Site : Dadanpatra)

Caste Category	*Occupation*			*Total*
	Fishing	*Cultivation*	*Other Occupation*	
Fishermen	24		2	26
	92.31		7.69	100.00
	85.71		13.33	41.27
Scheduled Castes/	4	5	1	10
Scheduled Tribes	40.00	50.00	10.00	100.00
	14.29	25.00	6.67	15.87
Other Castes		15	12	27
		55.56	44.44	100.00
		75.00	80.00	42.86
Total	28	20	15	63
	44.44	31.75	23.81	100.00
	100.00	100.00	100.00	100.00

Table 5.4 : Percentage Distribution of Respondents' Caste Category by House-Type

(Site : Dadanpatra)

Caste Category	*House-Type*			*Total*
	Thatched Hut	*Kuchcha*	*Pucca Having Thatched Roof/ Pucca*	
Fishermen	10 38.46 27.78	16 61.54 59.26		26 100.00 41.27
Scheduled Castes/ Scheduled Tribes	8 80.00 22.22	2 20.00 7.41		10 100.00 15.87
Other Castes	18 66.67 50.00	9 33.33 33.33		27 100.00 42.86
Total	36 57.14 100.00	27 42.86 100.00		63 100.00 100.00

Table 5.5 : Percentage Distribution of Respondents' Caste Category by Education

(Site : Dadanpatra)

Caste Category	*Education*				*Total*
	Illiterate	*Below Primary Level*	*Below Secondary Level*	*Beyond Secondary Level*	
Fishermen	6 23.08 30.00	13 50.00 46.43	6 23.08 46.15	1 3.85 50.00	26 100.00 41.27
Scheduled Castes/ Scheduled Tribes	6 60.00 30.00	3 30.00 10.71	1 10.00 7.69		10 100.00 15.87
Other Castes	8 29.63 40.00	12 44.44 42.86	6 22.22 46.15	1 3.70 50.00	27 100.00 42.86
Total	20 31.75 100.00	28 44.44 100.00	13 20.63 100.00	2 3.17 100.00	63 100.00 100.00

Table 5.6 : Percentage Distribution of Respondents' Caste Category by Women Help

(Site : Dadanpatra)

Caste Category	*Women Help*				
	Fish Feeding	*Farm Guarding*	*Marketing*	*No Help-House Wife only*	*No Response*
Fishermen	22 53.66	18 45.00			1 100.00
Scheduled Castes/ Scheduled Tribes	1 2.44	7 17.50	1 100.00	2 40.00	
Other Castes	18 43.90	15 37.50		3 60.00	
Total	41 100.00	40 100.00	1 100.00	5 100.00	1 100.00

It is quite heartening to find that most of the respondents are generally somewhat aware with the method and system followed in the co-operative societies. The findings given are based on an overall analysis of tables annexed with this report together with inferences drawn from individual or group discussions. The analyses indicates that only a very few are the members of fisherman co-operative society (Table 5.12) and about three-fourth of the respondents with males being more liberal are prepared to join a co-operative society having even other caste members (Table 5.10). However, all of them have favoured a society having the members only from within their village. Little more than two-third do not require any assistance from outside, they themselves can take care of the pond (Table 5.8). It is further pertinent to note that except a very few, all of them have felt that without the overall control of the fisheries official, they may not be able to run the society (Table 5.9). More than three-fourth have felt that they would like to operate the ponds for fish culture on individual basis (Table 5.7). The individuals will make their own efforts in taking care of their ponds, putting the seed, feeding and growing fish and catching them. However, for procuring the seed, feed and change of water and sale of fish, they will again need the help of fisheries officials. Their major help will be required for keeping accounts on distribution of shares and recovery of loans from them.

Table 5.7 : Percentage Distribution of Respondents' Caste Category by Attitude Towards Handling of Pond

(Site : Dadanpatra)

Caste Category	*Pond Handling*			*Total*
	Independently	*On Fixed Wages*	*On Coopera-tive Basis*	
Fishermen	17		9	26
	65.38		34.62	100.00
	34.69		90.00	41.27
Scheduled Castes/ Scheduled Tribes	8	2		10
	80.00	20.00		100.00
	16.33	50.00		15.87
Other Castes	24	2	1	27
	88.89	7.41	3.70	100.00
	48.98	50.00	10.00	42.86
Total	49	4	10	63
	77.78	6.35	15.87	100.00
	100.00	100.00	100.00	100.00

Table 5.8 : Percentage Distribution of Respondents' Caste Category by Attitude Towards Pond Care

(Site: Dadanpatra)

Caste Category	*Pond Care*		*Total*
	Can Take Himself/ Herself	*Assistance Required*	
Fishermen	19	7	26
	73.08	26.92	100.00
	47.50	30.43	41.27
Scheduled Castes/ Scheduled Tribes	2	8	10
	20.00	80.00	100.00
	5.00	34.78	15.87
Other Castes	19	8	27
	70.37	29.63	100.00
	47.50	34.78	42.86
Total	40	23	63
	63.49	36.51	100.00
	100.00	100.00	100.00

Table 5.9 : Percentage Distribution of Respondents' Caste Category by Attitude Towards Running of Co-operative Society

(Site: Dadanpatra)

Caste Category	*Running of Co-operative*			*Total*
	With Government / Fish. Dept. Help	*Without Government Help*	*No Response*	
Fishermen	23		3	26
	88.46		11.54	100.00
	39.66		60.00	41.27
Scheduled Castes/ Scheduled Tribes	10			10
	100.00			100.00
	17.24			15.87
Other Castes	25		2	27
	92.59		7.41	100.00
	43.10		40.00	42.86
Total	58		5	63
	92.06		7.94	100.00
	100.00		100.00	100.00

Table 5.10 : Percentage Distribution of Respondents' Caste Category by Joining of Co-operatives with Own Caste/Other Caste Members

(Site: Dadanpatra)

Caste Category	*Joining of Co-operatives*			*Total*
	With Own Caste Members	*With Other Caste Members*	*No Response*	
Fishermen	7	19		26
	26.92	73.08		100.00
	31.82	46.34		41.27
Scheduled Castes/ Scheduled Tribes	8	2		10
	80.00	20.00		100.00
	36.36	4.88		15.87
Other Castes	7	20		27
	25.93	74.07		100.00
	31.82	48.78		42.86
Total	22	41		63
	34.92	65.08		100.00
	100.00	100.00		100.00

Table 5.11 : Distribution of Respondents by Age

(Site: Dadanpatra)

Age Group	*Frequency (N)*	*Percentage (%)*
Below 20	—	—
20-40	37	58.73
Above 40	26	41.27
Total	63	100.00

N' is more because of multiple choice.

Table 5.12 : Distribution of Respondents by Present Co-operative Membership

(Site: Dadanpatra)

Member of Co-operative	*Frequency (N)*	*Percentage (%)*
Yes	3	4.76
No	60	95.24
Total	63	100.00

It is further a point to note that two-third of respondents have trusted fisheries officials and about one-third, their traditional caste members for handling the co-operative society money transactions. More so, only a very few have shown faith in the elected member for that purpose (Table 5.13) People are aware of the high market value of prawn and few even harvest this fish, though without any feeding or special care. A large majority of the respondents felt that the fish culture is to be developed in their individual ponds in a scientific manner, then, invariably the non-fishing families will also join this project (Table 5.14). On an overall basis. almost all of them have further felt that they will need training for soil and water quality checking as well as the scientific methods and techniques of prawn culture (Table 5.16). They also prefered that the government agencies excavate their ponds for them and then for its maintenance, they would need loans for one or two years. As regards women's role in brackish water fisheries development, most of the respondents felt that they would play a very important role in feeding of the fish and supervising and guarding the fish ponds. Another important role would be in keeping loan accounts (Table 5.6).

Co-operatives and Co-operation

There are nine Primary Fishermen Co-operative Societies (PFC)

registered with Central Co-operative Society (CFC), in the Digha-Contai area. Two of such societies, namely Ma-Naikali and Ma-Gange Fisherman Co-operative Society are in adjoining Ramnagar, Block-I. The CFC promotes, controls and provides finance to PFCs. Each of the PFCs has three groups of 10 fishermen each. Each of the group takes a mechanised boat and thus in a PFC there would be three such mechanised boats. These trawlers or launchers are used for deep sea fishing by group of fishermen. The fishermen after fish catch would return to fish landing locations at Digha or Shankarpur Mohana. The auction of the fish produce takes place and whosover bids highest, takes the produce through the commission agent. More so, the Fisheries Extension Officer on behalf of National Co-operative Development Corporation (NCDC) also acts as Chief Executive Officer for recovery of loan on the spot.

NCDC has already advanced loans for about 30 such mechanized boats and loans for another 100 is planned for this year. The NCDC loan scheme has been in operation for the past two years. As soon as the loan is fully repaid, the launch/trawler becomes group's property. However, it is important to note that the fish catch is distributed on a share basis, the cost of fuel and other maintenance expenses for the mechanised boat is also shared. The conflicts are locally resolved including damage suits of various kinds.

Table 5.13 : Distribution of Respondents by People Trusted

(Site: Dadanpatra)

People Trusted	*Frequency (N)*	*Percentage (%)*
Fisheries/Govt. Officials	41	65.08
Elected Members	2	3.17
Traditional Caste Leaders	19	30.16
None-Self Only	1	1.59
Secretary Co-operatives		
Total	63	100.00

Table 5.14 : Distribution of Respondents by Non-Fishing Family Participation in the Project

(Site: Dadanpatra)

Non-Fishing Family Like to Participate	*Frequency (N)*	*Percentage (%)*
Yes	50	79.37
No	13	20.63
Total	63	100.00

Table 5.15 : Distribution of Respondents by Pond Ownership

(Site: Dadanpatra)

Pond Ownership	*Frequency (N)*	*Percentage (%)*
No Pond	38	60.32
One Pond	20	31.75
Two Pond	4	6.35
Three Pond	1	1.59
Total	63	100.00

Table 5.16 : Distribution of Respondents by Type of Co-operatives Help

(Site: Dadanpatra)

Co-operatives Help Required	*Frequency (N)*	*Percentage (%)*
Loan	47	21.56
Technical Like Soil/Water Testing	51	23.39
Managerial	60	27.52
Extn. and Training	58	26.61
No Response	2	0.92
Total	218	100.00

'N' is more because of multiple choice.

Teghari-Purba Purushottampur is a registered co-operative society established 35 years ago for group farming. The society has allotted 545 acres of land to its 250 members, who belonged to Teghari and Purba Purushottampur villages. The land leased out to the society is located very close to Dakshin Purushottampur brackish water site. The society collected Rs. 2500 from each group of 10 members and got Rs. 5000 as grant from the government. The money was spend in constructing and strengthening of a sea dyke. As such then these 250 members started cultivating paddy on that 545 acres of land. At present, the membership has increased to 340. Bengal Salt Company, which was also leased out land, is located adjoining the society's land. The salt company claimed 250 acres of land as leased out to them and forced the society to vacate 250 acres of their land.

The matter is subjudiced in the court to law. Whatever is the status, the point to understand is that the co-operative society has been cultivating the land for the past several years. More than 150 families has permanently shifted to Dakshin Purushottampur and now these villagers have a very strong feeling that the government will take away their land and allot it to others for brackish water fish development scheme. This may have some ramification on the proposed project. Another point to understand is that the Secretary of the society is Postgraduate, young person who successfully ran the society except for the past few years, when it became defunct. However, the money is still collected from each group member regularly and court case is being fought. In spite of the fact that some people have even sub-let their share of land, the total acreage available and the number of original membership have undergone significant change and yet the balance land was mutually re-distributed and there does not seem to have arisen any major conflict between the society members.

For promoting brackish water fisheries, the Brackish Water Fisheries Development Agency (BFDA) arranges training for private prawn culture development and seed collectors. In the study villages, some people were found to have taken training and only very few were cultivating prawn on their farm. However, in Purba Purushottampur, a few farmers took back loan and excavated ponds which were in Dakshin Purushottampur. One year they got good profit, paid back loan amount and then the cyclone disturbed their ponds and no repayments of loans thereafter, were made.

But in another case, the Contai Co-operative Bank, Ramnagar had advanced a loan of Rs 2,00,000 to a group of fisherman, who had also put in Rs. 1,00,000 for the purchase of a mechanised boat. Though these fisherman are from affluent urban group, the repayment of loan, as per Bank Manager, is regular.

For promoting the prawn, the fisheries department has an experimental hatchery, which is located at Digha. Adjoining it is United Nations Development Programme (UNDP) Alampur Brackish Water Fisheries Project on 40 hectares area. All these ponds have already been excavated, and are of one hectare size. State Fisheries Development Corporation (SFDC) is likely to manage the ponds while the technical support for fish growth would be given by the BFDA, SFDC besides fish farm management and its operation, will look after the marketing aspect also. Another scheme in this area is a Central Assistance Scheme (Fifty-Fifty basis). It is opposite Shankarpur having a land of 123 hectares with pond size of 0.5 hectare each. 200 such ponds then are proposed to be given to poor people. The terms and conditions for allotment/lease, etc. are yet to be finalised. Yet other scheme on UNDP pattern is proposed near Purba Mukundapur on 20 ha. land area. The engineering survey is completed. The World Bank proposed scheme is located at Dakshin Purushottampur. The micro level engineering and Biological Survey by MPEDA have already been completed.

All this goes to show that a good infrastructure for brackish water fisheries development is already existing in this area and people though have not much experience either on individual or co-operative basis of prawn cultivation, but they collect prawn seed, sell them at good rates. On marine fishing aspect, people have good experience of taking group loans for arranging trawler or launch undertake fish catch, distribute the share and repay the loan at the fish landing spot itself.

BRACKISH WATER SITE : DIGHA

Digha — an attractive tourist centre forms a part of Ramnagar Block-I of District Medinipur. The Block Headquarters is seven km away on Digha-Contai road. Digha brackish water site is about four km north-east of Digha town. The nearest village to this site is Purba Mukundapur. The site is located inside the sea dyke and is approachable by jeep upto some distance through sea dyke. It is also approachable by metalled and Mooharam Road up to Purba Mukundapur. From Purba Mukundapur, one has to walk about a kilometre to reach the site. Within a radius of five km of this site, there are seven revenue villages. These villages are approachable by road and walk.

Profile of the Villages

The pattern of village settlement in Digha brackish water site is generally small scattered and stretched along the road (see Map 5.4 and 5.5). A total of 15 settlements excluding two *khatti settlements* (seasonal camps) comprising of seven revenue villages are situated within a radius of five km from the Digha brackish water site (see Map 5.6). Purba Mukundapur is a medium size village with a population of about 2,000 scattered over four hamlets, namely Dasanpur Asanpur, Khas and Ekdal. This revenue village falls under the Gram Panchayat of Podima-II. At Digha Muhana or Shankarpur Muhana (except during the rainy season), around 250 fisherman camp. For some period, they have their camps at Shankarpur and then towards winter they shift to Digha Muhana. A Muhana is a fish landing location and all major fish marketing takes place at Muhana only. Digha barckish water site villages, being nearer of Digha town, have the added advantage of doing labour work and rickshaw pulling in the town. They are also involved in the preparation of various attractive items of sea weeds which they sell to the tourists at Digha.

Out of seven selected villages, four are part of Ramnagar Police Station and remaining three Digha Police Station. Digha Police Station has only 32 villages, while Ramnagar has a total of 287 villages. However, site villages of both the Police Stations (PS's) fall under Ramnagar Block-I. One-third of the villages in both PS's are small having less than 500 population and so are almost all the site except Purba Mukundapur — a medium size village.

About two-third of the area of the Ramnagar and Digha PS's is cultivable to total area. The percentage of irrigated area to total cultivable area, however, is very low (seven per cent). The situation in site villages is however, better as some of the farmers use their diesel pump sets and take two crops of paddy in a year. The site villages in terms of approach road are slightly better than other PS's villages. In the site villages, the availability of electricity and medical amenities is relatively poor. But on the whole, drinking water and educational facilities are generally better. Agriculture is the predominant occupation in the site villages. Equally or even more important occupation is fishing. Betal leaves and salt making, labour work in the town area are other important means of livelihood of the site villages. The availability of high lands with alluvial soil together with the sea dyke for its protection, availability of diesel pumps for agriculture and nearness to the sea coast as well as proximity to an important tourist town and frequent availability of public buses and pedal rickshaw indicate that the site villages have a relatively better developmental infrastructure and are relatively better off.

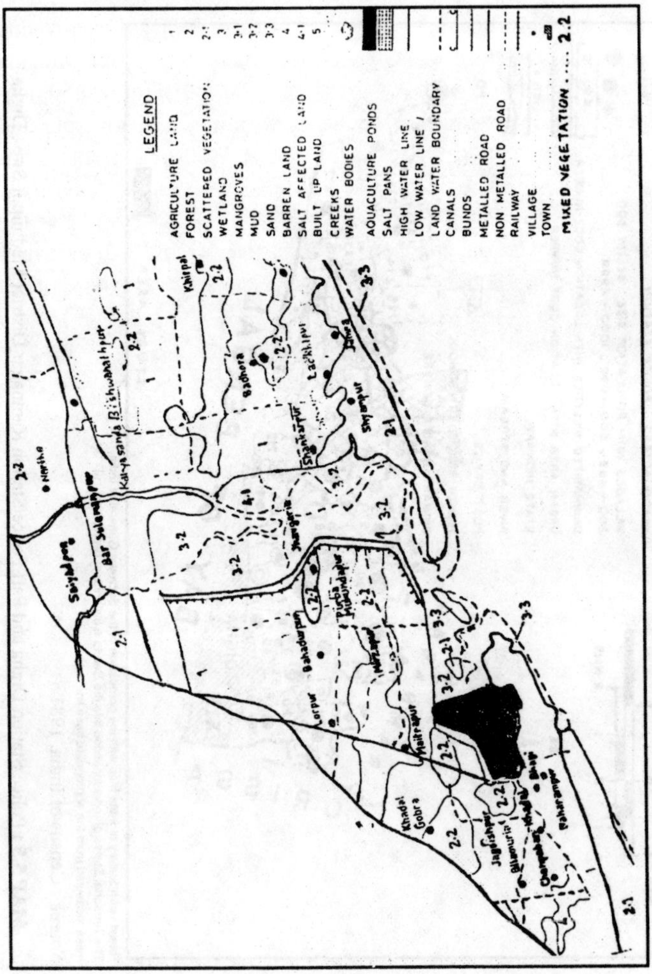

Source : Remote Sensing Applications Group, ISRO, Ahmedabad, 1990.

MAP 5.4 : Landuse Map Site : Digha

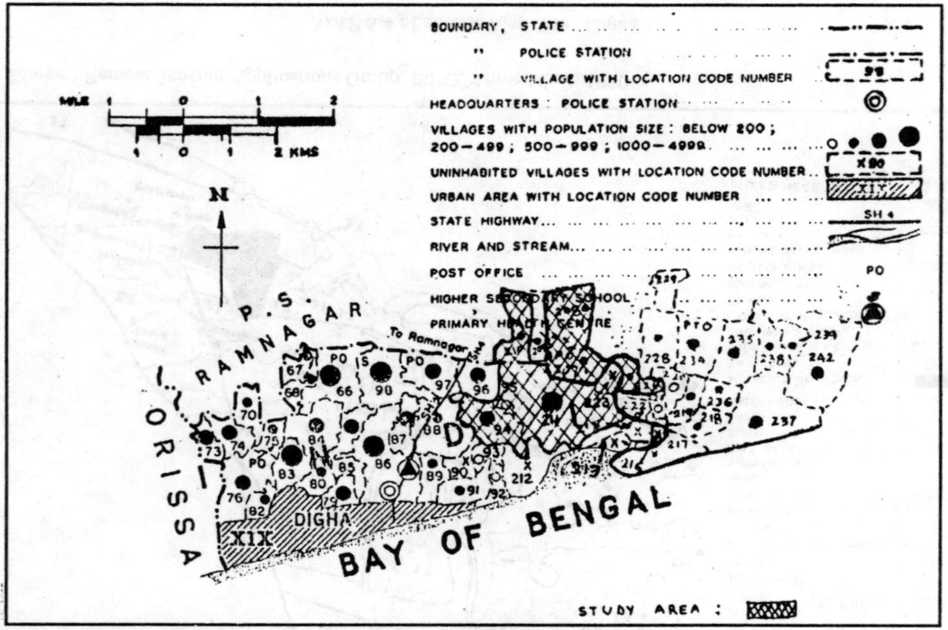

Based upon Survey of India Map with the permission of the Surveyor General of India.
The Territorial States of India extended into the Sea to a distance of twelve nautical miles measured from the appropriate base line

Source : Census of India, 1981.

MAP 5.5 : Police Station Digha and Part Police Station Ramnagar District Medinipur Site : Digha

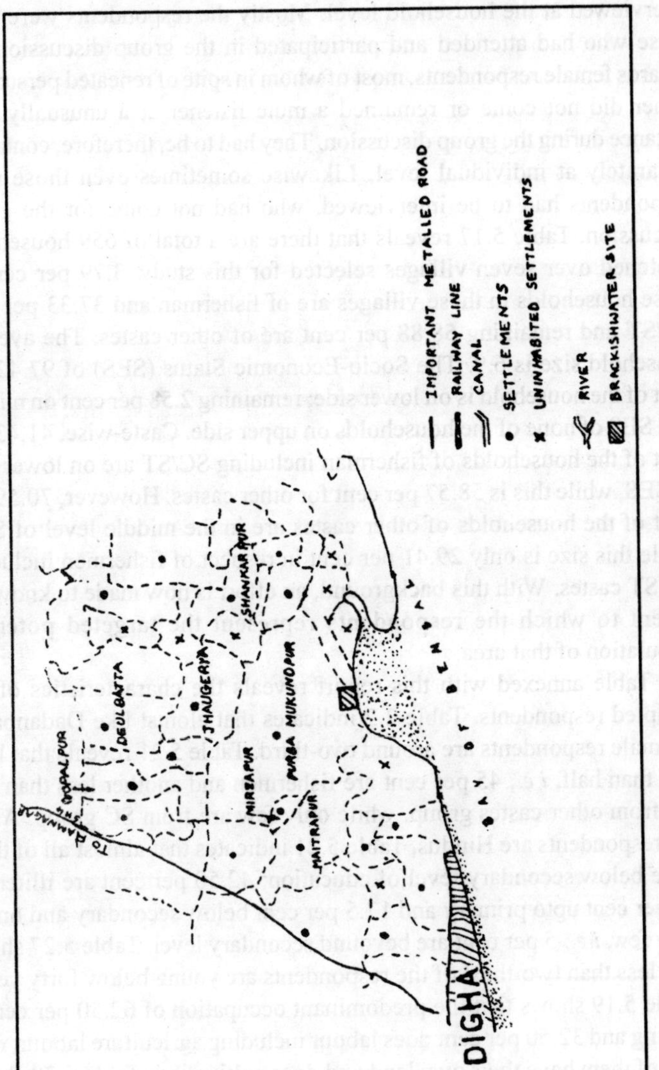

MAP 5.6 : Map showing villages visited at Digha Brackish Water Site

Analysis of Individual Responses

For the present study forty respondents spread over seven villages were interviewed at the household level. Mostly the respondents were from those who had attended and participated in the group discussion. As regards female respondents, most of whom in spite of repeated persuasion either did not come or remained a mute listener at a unusually long distance during the group discussion. They had to be, therefore, contacted separately at individual level. Likewise sometimes even those male respondents had to be interviewed, who had not come for the group discussion. Table 5.17 reveals that there are a total of 659 households stretched over seven villages selected for this study. 3.79 per cent of these households in these villages are of fisherman and 37.33 per cent SC/ST and remaining 58.88 per cent are of other castes. The average household size is 6.9. The Socio-Economic Status (SES) of 97.42 per cent of the household is on lower side; remaining 2.58 per cent on middle and SES of none of the households on upper side. Caste-wise, 41.43 per cent of the households of fisherman including SC/ST are on lower side of SES, while this is 58.57 per cent for other castes. However, 70.59 per cent of the households of other castes are in the middle level of SES, while this size is only 29.41 per cent in respect of fishermen including SC/ST castes. With this background, an effort is now made to know the extent to which the respondents represent the targeted potential population of that area.

Table annexed with this report reveals the characteristics of the sampled respondents. Table 2.2 indicates that alomst like Dadanpatra, the male respondents are around two-third. Table 5.21 reveals that little less than half, *i.e.*, 45 per cent are fishermen and another less than half are from other castes group, while only few are from SC group. All of the respondents are Hindus. Table 5.21 indicates that almost all of them have below secondary level of education; 42.50 per cent are illiterate; 35 per cent upto primary and 17.5 per cent below secondary and only a very few, *i.e.*, 5 per cent are beyound secondary level. Table 5.27 shows that less than two-third of the respondents are young below forty years. Table 5.19 shows that the predominant occupation of 62.50 per cent is fishing and 32.50 per cent does labour including agriculture labour; only two of them have their own land and does cultivation. Table 5.20 shows that 65 per cent of them stay in thatched hut and the remaining in *kuchcha* houses, most of whom have roof of country made tiles. Table 5.31 shows that almost three-fourth of them have no pond and only one-fourth have

Table 5.17 : Distribution of Caste Category by Households, Population and Socio-Economic Status in Study Villages

(Site : Digha)

Caste Category	*Number of Household*	*Total Population*	*Household Size*	*Socio-Economic Status*		
				Lower	*Middle*	*Upper*
Fishermen	25	300	12.0	25		
	100.00			100.00		
	3.79	6.63		3.89		
Scheduled Castes/	246	1850	7.5	241	5	
Scheduled Tribes	100.00			97.97	2.03	
	37.33	40.88		37.54	29.41	
Other Castes	388	2375	6.1	376	12	
	100.00			96.91	3.09	
	58.88	52.49		58.57	70.59	
Total	659	4525	6.9	642	17	
	100.00			97.42	2.58	
	100.00	100.00		100.00	100.00	

Table 5.18 : Percentage Distribution of Respondents' Occupation by Annual Income

(Site: Digha)

Occupation	*Annual Income*			*Total*
	< 3000	*3000-5000*	*Above 5000*	
Fishing	14	9	2	25
	56.00	36.00	8.00	100.00
	93.33	47.37	33.33	62.50
Cultivation	1	1		2
	50.00	50.00		100.00
	6.67	5.26		5.00
Other Occupation		9	4	13
		69.23	30.77	100.00
		47.37	66.67	32.50
Total	15	19	6	40
	37.50	47.50	15.00	100.00
	100.00	100.00	100.00	100.00

Table 5.19 : Percentage Distribution of Respondents' Caste Category by Occupation

(Site : Digha)

Caste Category	*Occupation*			*Total*
	Fishing	*Cultivation*	*Other Occupation*	
Fishermen	18 100.00 72.00			18 100.00 45.00
Scheduled Castes/ Scheduled Tribes	3 50.00 12.00		3 50.00 23.08	6 100.00 15.00
Other Castes	4 25.00 16.00	2 12.50 100.00	10 62.50 76.92	16 100.00 40.00
Total	25 62.50 100.00	2 5.00 100.00	13 32.50 100.00	40 100.00 100.00

Table 5.20 : Percentage Distribution of Respondents' Caste Category by House-Type

(Site : Digha)

Caste Category	*House-Type*			*Total*
	Thatched Hut	*Kuchcha*	*Pucca Having Thatched Roof/ Pucca*	
Fishermen	11 61.11 42.31	6 33.33 46.15	1 5.56 100.00	18 100.00 45.00
Scheduled Castes/ Scheduled Tribes	6 100.00 23.08			6 100.00 15.00
Other Castes	9 56.25 34.62	7 43.75 53.85		16 100.00 40.00
Total	26 65.00 100.00	13 32.50 100.00	1 2.50 100.00	40 100.00 100.00

Table 5.21 : Percentage Distribution of Respondents' Caste Category by Education

(Site : Digha)

Caste Category	*Education*				*Total*
	Illiterate	*Below Primary Level*	*Below Secondary Level*	*Beyond Secondary Level*	
Fishermen	10 55.56 58.82	3 16.67 21.43	4 22.22 57.14	1 5.56 50.00	18 100.00 45.00
Scheduled Castes/ Scheduled Tribes		6 100.00 42.86			6 100.00 15.00
Other Castes	7 43.75 41.18	5 31.25 35.71	3 18.75 42.86	1 6.25 50.00	16 100.00 40.00
Total	17 42.50 100.00	14 35.00 100.00	7 17.50 100.00	2 5.00 100.00	40 100.00 100.00

Table 5.22 : Percentage Distribution of Respondents' Caste Category by Women Help

(Site : Digha)

Caste Category	*Women Help*				
	Fish Feeding	*Farm Guarding*	*Marketing*	*No Help-House Wife only*	*No Response*
Fishermen	9 32.14	9 52.94	1 100.00	1 33.33	
Scheduled Castes/ Scheduled Tribes	5 17.86	3 17.65			
Other Castes	14 50.00	5 29.41		2 66.67	
Total	28 100.00	17 100.00	1 100.00	3 100.00	

'N' is more because of multiple choice.

Table 5.23 : Percentage Distribution of Respondents' Caste Category by Attitude Towards Handling of Pond

(Site : Digha)

Caste Category	*Pond Handling*			
	Independently	*On Fixed Wages*	*On Co-operative Basis*	*Total*
Fishermen	15 83.33 41.67	2 11.11 100.00	1 5.56 50.00	18 100.00 45.00
Scheduled Castes/ Scheduled Tribes	6 100.00 16.67			6 100.00 15.00
Other Castes	15 93.75 41.67		1 6.25 50.00	16 100.00 40.00
Total	36 90.00 100.00	2 5.00 100.00	2 5.00 100.00	40 100.00 100.00

Table 5.24 : Percentage Distribution of Respondents' Caste Category by Attitude Towards Pond Care

(Site : Digha)

Caste Category	*Pond Care*		
	Can Take Himself/ Herself	*Assistance Required*	*Total*
Fishermen	8 44.44 33.33	10 55.56 62.50	18 100.00 45.00
Scheduled Castes/ Scheduled Tribes	3 50.00 12.50	3 50.00 18.75	6 100.00 15.00
Other Castes	13 81.25 54.17	3 18.75 18.75	16 100.00 40.00
Total	24 60.00 100.00	16 40.00 100.00	40 100.00 100.00

ponds which are generally small and situated in front of their houses. Table 3.3 reveals that a great majority of them also do not have land and only one-fourth have land, which is below one acre. Table 3.4 shows that majority of them have no cattle and only less than even one-fourth have some cattle like cow, duck, poultry, etc. Table 3.7 shows that most of them have never taken any bank loan though a few have the experience of taking such a loan. Table 5.18 indicates that income-wise, most of them are poor; a great majority of them have annual income below Rs. 5,000 only a very few earn more than Rs. 5,000. It was elsewhere observed that almost two-third have poor mass-media exposure and the remaining one-third have a relatively higher mass-media exposure. All the above goes to indicate that generally the respondents belong to varied groups of the fishermen, Scheduled Castes and other castes categories and with lesser number of female respondents. This is because of the social setting of the area where substantially large number of people belong to other castes and yet undertake fishing as one of their most important occupation. The role of women is generally limited to household activities and certainly have no role in deep sea fishing or selling it at fish landing locations. The overall analysis of data of respondents indicate that most of the respondents are poor, staying in thatched huts and have no ponds or lands of their own, have poor mass-media exposure. All this goes to indicate that the sampled respondents are generally representative of the targeted poor population of the brackish water project.

An overall analysis of tables annexed with this report together with personal inferences drawn from various individual level and group based discussions indicate that a large majority of them are not the members of any fisherman co-operative society (see Table 5.28). However, it is very important to know that the respondents are generally aware with the system and method followed in co-operative societies. They were not much aware with technicalities of elections, registration procedure, etc., but certainly they had fairly good understanding on the functioning of primary fishermen and agricultural credit co-operative societies. Most of the respondents including the women were quite liberal and prepared to join co-operative society having other caste members (see Table 5.26). However, invariably all of them preferred to join such a society which was small and village based. Each of them was against joining a larger co-operative society. They felt that most of these large societies are only meant for giving advantage to economically better and politically powerful persons.

Table 5.25 : Percentage Distribution of Respondents' Caste Category by Attitude Towards Running of Co-operative Society

(Site: Digha)

Caste Category	*Running of Co-operative* With Government / Fish. Dept. Help	*Without Government Help*	*No Response*	*Total*
Fishermen	15 83.33 40.54	1 5.56 100.00	2 11.11 100.00	18 100.00 45.00
Scheduled Castes/ Scheduled Tribes	6 100.00 16.22			6 100.00 15.00
Other Castes	16 100.00 43.24			16 100.00 40.00
Total	37 92.50 100.00	1 2.50 100.00	2 5.00 100.00	40 100.00 100.00

Table 5.26 : Percentage Distribution of Respondents' Caste Category by Joining of Co-operatives with Own Caste/Other Caste Members

(Site : Digha)

Caste Category	*Joining of Co-operatives* With Own Caste Members	*With Other Caste Members*	*No Response*	*Total*
Fishermen	5 27.78 62.50	13 72.22 40.63		18 100.00 45.00
Scheduled Castes/ Scheduled Tribes	ERR ERR	6 100.00 18.75		6 100.00 15.00
Other Castes	3 18.75 37.50	13 81.25 40.63		16 100.00 40.00
Total	8 20.00 100.00	32 80.00 100.00		40 100.00 100.00

Table 5.27 : Distribution of Respondents by Age

(Site: Digha)

Age Group	*Frequency (N)*	*Percentage (%)*
Below 20	1	2.50
20-40	24	60.00
Above 40	15	37.50
Total	40	100.00

Table 5.28 : Distribution of Respondents by Present Co-operative Membership

(Site: Digha)

Member of Co-operative	*Frequency (N)*	*Percentage (%)*
Yes	5	12.50
No	35	87.50
Total	40	100.00

Table 5.29 : Distribution of Respondents by People Trusted

(Site: Digha)

People Trusted	*Frequency (N)*	*Percentage (%)*
Fisheries/Govt. Officials	27	67.50
Elected Members	4	10.00
Traditional Caste Leaders	7	17.50
None-Self Only	1	2.50
Secretary Co-operatives	1	2.50
Total	40	100.00

'N' is more because of multiple choice.

Table 5.30 : Distribution of Respondents by Non-Fishing Family Participation in the Project

(Site : Digha)

Non-Fishing Family Like to Participate	*Frequency (N)*	*Percentage (%)*
Yes	25	62.50
No	15	37.50
Total	40	100.00

Table 5.31 : Distribution of Respondents by Pond Ownership

(Site : Digha)

Pond Ownership	*Frequency (N)*	*Percentage (%)*
No Pond	30	75.00
One Pond	10	25.00
Two Pond	—	—
Three Pond	—	—
Total	40	100.00

Table 5.32 : Distribution of Respondents by Type of Co-operatives Help

(Site : Digha)

Co-operatives Help Required	*Frequency (N)*	*Percentage (%)*
Loan	25	21.55
Technical Like Soil/Water Testing	18	15.52
Managerial	40	34.48
Extn. and Training	32	27.59
No Response	1	0.86
Total	116	100.00

'N' is more because of multiple choice.

Further, much more than half of them felt confident to take care of the pond themselves and the remaining required the help of some other family members (Table 5.24). Invariably, without the intervention of fisheries officials, it was impossible for the them to run even the small village based societies (Table 5.25). The fisheries officials' help was required not only in identification of seeds, conducting of water and soil quality tests but more importantly in keeping proper accounts, marketing the fish and distributing the share to the members. Table 5.23 shows that as regards pond handling, almost all of them felt that the ownership of pond should be individual based and so the efforts for maintenance of pond, fish feeding, putting manure, etc. However, efforts for changing of water, arranging of seed, fish feed and marketing should be made by the village based co-operative society under overall control of the fisheries official. The most trusted person for handling the money transactions is the Fisheries Extension officials. It is quite pertinent to note that the elected members are not much trusted for this and the traditional caste based leaders have been rated to be having poor exposure of the fast changing world (Table 5.29). Since most of the people are aware with the high market value of prawn, they have felt that even the non-fishing families would happily undertake and actively participate in any of the aspects of brackish water fisheries development project (Table 5.30).

Besides bank loan, the requirement of most of the respondents for cultivating fish in brackish water ponds, runs around the help for managing the fish production and its sale. For purposes of growing fish, they were asking for providing training to help them identify diseases in the fish, water and soil quality checks and other techniques for cultivating the prawn fish (Table 5.32). As regards women's help, the respondents generally felt that their role would be limited to household or maximum fish feeding or guarding the ponds and they have further felt that the women folk is good in keeping loan accounts (Table 5.22).

Co-operatives and Co-operation

There are nine Primary Fishermen Co-operative Societies (PFC) registered with Central Co-operative Society (CFC), in the Digha-Contai area. Two of such societies, namely Ma-Naikali and Ma-Gange Fisherman Co-operative Society are in Ramnagar Block-I itself. The CFC promotes, controls and provides finance to PFC's. Each of the

PFC's has three groups of 10 fishermen each. Each of the group takes a mechanised boat and thus in a PFC there would be three such mechanised boats. These trawlers or launchers are used for deep sea fishing by group of fishermen. The fishermen after fish catch would return to fish landing locations at Digha or Shankarpur Mohana. The auction of the fish produce takes place and whosoever bids highest, takes the produce but through the commission agent. More so, the Fisheries Extension Officer on behalf of National Co-operative Development Corporation and acts as Cheif Executive officer for recovery of loan on the spot.

NCDC has already advanced loans for about 30 such mechanized boats and loans for another 100 is planned for this year. The NCDC loan scheme has been in operation for the past two years. As soon as the loan is fully repaid, the launch/trawler becomes group's property. However, it is important to note that the fish catch is distributed on a share basis, the cost of fuel and other maintenance expenses for the mechanised boat is also shared. The conflicts are locally resolved including damage suits of various kinds.

For inland fisheries development, there are a total of 21 Primary Fisherman Co-operative Societies. One of them Podima Anchal Primary Fisherman Co-operative Society, Purba Mukundapur, has membership of 120, out of which 80 are from Purba Mukundapur village itself. The PFC's get the loans for fish net, and fish culture between Rs. 10,000 to Rs. 20,000. The loan is also taken for fish culture in Gram Panchayat leased in ponds. The posting of a Supervisor for collection of loan installment on the sale spot has considerably helped in improving the loan repayment situation in the area. Yet in other case the Contai Co-operative Bank, Ramnagar had advanced a loan of Rs. 2,00,000 to a group of fisherman, who had also put in Rs. 1,00,000 for the purchase of a mechanised boat. Though these fishermen are from affluent urban group, the repayment of loan, as per Bank Manager, is regular.

Alankarpur Co-operative Society is a registered Co-operative Credit Society and is in operation for the past 10 years. Each member has minimum share of Rs. 1,000. Any member in need of money applies for loan. On judging the suitability it is sanctioned. The members can take loan but have to pay little higher interest rate. In case of non-payment, one member of the village, who acts as Director, presses the loanee hard for recovery and there does not seem problem of repayment. All the money is managed through the bank. Jhaugerya, Purba Mukundapur has another agricultural credit co-operative society and most of the families of Purba Mukundapur are its members. No major conflicts of repayment of loan was found.

In Purba Mukundapur, there is a 21 member *Gram* or Village Committee which solves any social conflicts. If they unable to settle only then the matter is referred to Gram Panchayat (Village Council). Generally the matter is resolved at the village level itself. Such comittees are operational in all the villages of this area.

Very close to Purba Mukundapur, Digha Fishermen and Fish Trading Association is located at Digha Mohana. The association has its office at Mohana and collects Rs. 21 per month per mechanised boat which are 200-250 in number. The association spends the money in repairing approach road from Digha to Mohana, arranging a diesel sale depot. Most important among the activity is the money spent on 'Pooja' festival.

There are also few failure cases in the area. As per the Block Development Officer, Ramnagar-I, a few years ago, 12 fishermen got a co-operative society registered took loan and actually purchased a launch. Everybody's catch was sold to the commission agent. All the members of the group assured that the money is collected on behalf of all of them by the commission agent as their share of repayment of loan. This process continued for very long and one day the Secretary/President varried away all the money from the commission agent. He did not repay the loan and pocketed all the money. One of the reason as the B.D.O. felt for malfunctioning of the co-operative society is related to the fact that the leaders are generally much more shrewd, clever as compared to its members.

Under Integrated Rural Development Programme (IRDP), ponds were excavated at Alankarpur (near site village) and given to poor on individual basis. But for want of proper management system and availability of loans, these ponds are not effectively operated by these individuals. In yet other individual brackish water fisheries development scheme, a few farmers of Shankarpur took the loan Contai Bank, Ramnagar. However, before the production could start, cyclone devastated the ponds and these loanees could not return the loan. The bank has now a condition that unless these ponds have an insurance cover, the loans will not given.

There is still another failure case of a Co-operative Society in Purba Purushottampur, a Primary Fishermen Co-operative Society having 15 members from this village initially were engaged in taking group auction of fishing at creeks. Subsequently, the society took loan from NCDC for purchase of launch and nets. The launch sank while at sea and the Secretary of the society sold nets and kept the money.

6

ORISSA

In this Chapter two brackish water sites namely, Jagatjore and Panaspada have been described and for each site, separate socio-economic analyses has been presented. All efforts have been made to discuss the major issues concerning the semi-intensive brackish water shrimp farming.

BRACKISH WATER SITE : JAGATJORE

Although neither the Fisheries Extension Officer nor the officials of Orissa Maritime Chilka Area Development Agency (OMCADA), a government sponsored autonomous body operating both in the area of forestry and fisheries development, have any idea of the exact location of the site but according to their guess it is close to the Jagatjore market (see Map 6.1) . This plot of 465 hectares belongs to the Revenue Department of Orissa Government and was given to OMCADA 10 years ago for coconut plantation. In the nearby areas, OMCADA has another about 1000 hectare of land which they have used mainly for forestry and another 190 hectare of land where they have developed brackish water fisheries. Sixty hectares of land has been leased out to a private party for prawn culture.

For the last 10 years, OMCADA could not develop this area except 66 hectares, mainly, because the land is low lying and has a tidal creek inside which allows water to stagnate. But OMCADA seems to be in no mood to return this land to revenue department which has agreed to give this land to fisheries department for World Bank Project. The negotiation for transfer is in progress. The general concern in the mind of the OMCADA officials is about the future of nearly 50 labourers presently working for the last 10 years for this land as daily wage labourers.

Out of 320 hectares of identified farm area, an estimated area of 65 per cent *i.e.* 200 hectare will be the net water area and the rest will be for embankment.

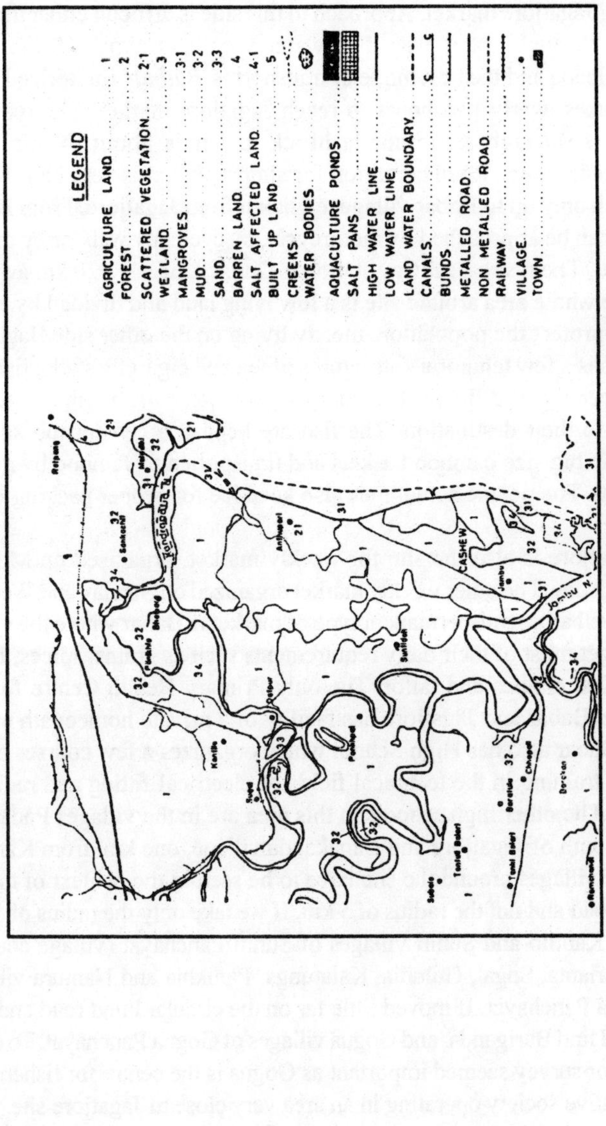

Source : Remote Sensing Applications Group, ISRO, Ahmedabad, 1990.

Map 6.1 : Landuse Map Site : Jagatjore

Profile of the Villages

The proposed site for Brackish Water Fisheries Development is part of Kalatunga revenue village under Gogua Panchayat of Mahakalpada block in Cuttack district (see Map 6.2). It is located by the side of the creek reaching Jagatjore market. Approach to this side is difficult either through road or through water way. For water way it is about 10 km from Mahakalpada and the best mode of transport is '*Butbuti*' (motorized boat) which takes nearly two hours to reach Jagatjore market. The road approach is through Kendrapada block and it is about 30 km from Pattamundai town. Road condition for almost six km is very bad. A jeep can reach only up to Babar village, eight km from Jagatjore. From Babar, the site can be approached by motorcycle, by cycle, by walk or by motorized boat. The nearest railway station is Cuttack, nearly 120 km away.

The whole area around site is a low lying land and divided by a high bund to protect the population, mostly living on the other side. Jagatjore market has a few temporary structures of tea and cigarette stalls, fish collection centres of different fish companies which collect fish and send it onward to their destination. The fish are kept in a cement tub and are packed in big size bamboo baskets and transported to Jamboo by motorized boat. The people sometimes also keep ice for proper packing of the fish.

Jagatjore is also the site for weekly market, organised on Monday and Thursday. The other weekly market organized on Sunday and Wednesday as well as a semi-permanent type of market is Babar where the villagers can get most of their daily requirements such as grains, spices, fruits, vegetables, clothes and tailor. The only Primary Health Centre for this area is in Babar and Jagatjore has facility of a private homeopath practitioner. Babar also has High School which organizes a few courses of vocational training in the technical fields of electrical fitting and radio repairing. The other high schools in this area are in the villages Padapalli, six km south of Jagatjore and Banakanda village, one km. from Kantilo.

The villages around the site have to be seen in the context of the approach road and not the radius of 5 km. If we take only the radius of 5 km, we find Kantilo and Suniti villages of Suniti Panchayat (village council) and Kharianta, Sugal, Guludia, Kalatunga, Panikhia and Nanjura villages of Gogua Panchayat. If moved little far on the circular bund road and then we could find Bariganda, and Gogua villages of Gogua Panchayat. To cover Gogua for survey seemed important as Gogua is the centre for fisherman's co-operative society operating in an area very close to Jagatjore site.

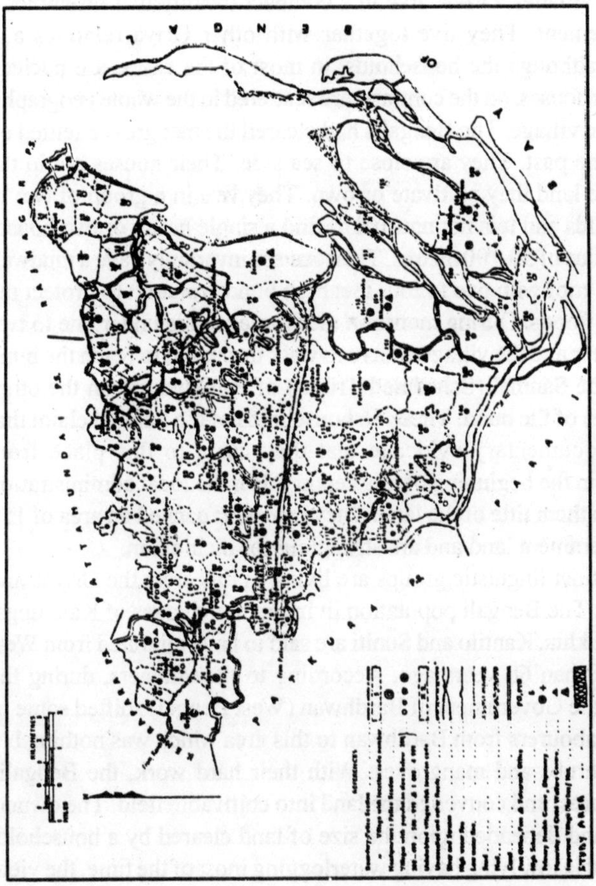

Based upon Survey of India Map with the permission of the Surveyor General of India.
The Territorial States of India extended into the Sea to a distance of twelve nautical miles measured from the appropriate base line

Source : Census of India, 1981.

Map 6.2 : Mahakalapada P. S. District : Cuttack (Orissa) Site : Jagatjore

The villages as shown in the Map 6.3 and as listed in the village list (see Annexure 4) fall mainly under the Gogua Panchayat. Kantilo and Suniti villages are part of Suniti Panchayat. The villages of Gogua Panchayat are mainly by the side of or on the inner side of the circular bund road.

There is a marked difference in the settlement pattern of the Oriya and Bengali villages. Oriyas live in a compact locality that resembles a village settlement. They live together with other Oriya relatives and neighbours, although the households, in most of the cases, are nuclear. The Bengalis houses, on the contrary, are scattered in the whole geographical area of the village. The Bengalis had cleared the mangroves settled on the land in the past. They are close to sea side. Their houses are in the middle of the land they cultivate or own. They live in a group of two to five households and it is not unusual to find a single house or two households in a radius of 50-100 metres. Their houses invariably have a mudwall and thatched roof with two to four feet high bund around it to protect the houses from floods. During monsoon the whole area is under one to two feet deep water and the villagers have to walk through it to reach the bund road. Only the Santhal (Scheduled Tribes) of Panikhia are on the other side (sea side) of the bund. These 36 households of the Santhal claim that they have documentary evidence that they came to this place from Mayurbhanj in the beginning of this century. But the local administration has not given them title of the land. They have encroached an area of 151 areas of government land and are struggling for their claim.

Two distinct linguistic groups are living in this area; the Oriyas and the Bengalis. The Bengali population living in the villages of Kalatunga, Guludia, Panikhia, Kantilo and Suniti are said to have migrated from West Bengal more than 60 years ago. According to the villagers, during the British rule, the Government of Bardhwan (West Bengal) shifted some of the Bengali labourers from Bardhwan to this area which was nothing but a forest of shrubs and mangroves. With their hard work, the Bengalis cleared the forest and converted the land into cultivable field. The ownership of the land depended upon the size of land cleared by a household. Since the land is low lying having waterlogging most of the time, the yield is very low so nobody bothered about its entitlement. Most of these Bengali households have got one to two acres of land on their name but in terms of their income, they mostly depend upon the daily wage earning as labourers. Since cultivation is not a profitable proposition, those having large acres of land in the neighbouring villages do not generally engage the labourers. Only a few villagers from Gogua, Nanjura, Sugal and other villages

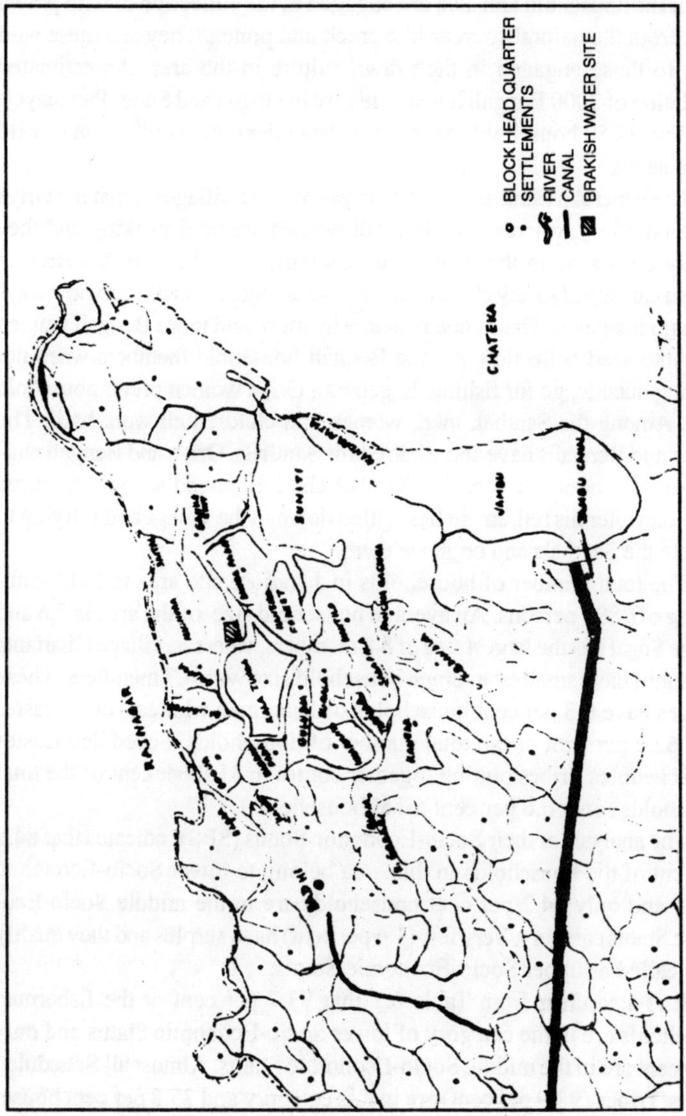

Map 6.3 : Map showing villages visited at Jagatjore Brackish Water Site

engage them for transplantation and harvesting. The Bengali men walk up to a distance of 10-12 km to get some employment. Bengali women take care of the household and also are engaged in the collection of wild prawn seeds from the natural sources like creek and ponds. They sell these wild seeds to those engaged in the prawn culture in the area. An estimated population of 2000 Bengali households are in Gogua and Suniti Panchayats and almost 250 households are mainly dependent upon collection of wild prawn seeds.

The general feeling among the Oriyas of these villages is that the Oriya women are lazy, whereas, the Bengali women are hard working and they equally contribute to the household economy. They believe that Bengali women can adjust to any circumstances and are more receptive to new ways of economic gains. This is one reason why more and more Bengali women enter into seed collection. All the Bengali household members when not working outside, go for fishing. In general, Oriya women prefer household work. Among the Santhal, men, women and children all work hard. The Oriyas and Bengalis have some hatred for Santhals. Oriya and Bengali children go to school but very few Santhal children attend school. A retired army man after his retirement has settled down in the village and is trying to educate the Santhals and organize them.

The total number of households in Jagatjore site area is 1414 comprising of 7895 persons. An average household size of the area is 5.6 and village Sugal has the largest size of 8.8 members, whereas, village Kharianta and Suniti have smallest average household size with 4.7 members. These villages have 6.8 per cent households of fisherman whereas, other castes form 62.2 per cent of the total number of households. Scheduled Castes and Schedules Tribes putting together comes to 31.0 per cent of the total households (only 6.6 per cent tribal households).

The analysis of their Socio-Economic Status (SES) indicates that 84.2 per cent of the households in this area belong to lower Socio-Econoimc Status and only 14.2 per cent households are in the middle Socio-Economic Status category. Very few (1.6 per cent) have surplus and they qualify to be called as upper Socio-Economic Status.

It is also clear from Table 6.1 that 93.7 per cent of the fisherman households are in the category of lower Socio-Economic Status and only 6 per cent are in the middle Socio-Economic Status. Almost all Scheduled Castes/Tribes (98.9 per cent) are in this category and 75.8 per cent households belong to lower Socio-Economic Status followed by 21.6 per cent in middle Socio-Economic Status and only 2.6 per cent who are in the cateogry of upper Socio-Economic Status.

Table 6.1 : Distribution of Caste Category by Households, Population and Socio-Economic Status in Study Villages

(Site : Jagatjore)

Caste Category	*Number of Household*	*Total Population*	*Household Size*	*Socio-Economic Status*		
				Lower	*Middle*	*Upper*
Fishermen	96 (6.8)	560 (7.1)	5.8	90 (9.6) (93.7)	6 (3.0) (6.3)	
Scheduled Castes/ Scheduled Tribes	439 (31.0)	2145 (27.2)	4.9	434 (36.4) (98.9)	5 (2.5) (1.1)	
Other Castes	879 (62.2)	5190 (65.7)	5.9	666 (56.0) (75.8)	190 (94.5) (21.6)	23 (100) (2.6)
Total	1414 (100.0)	7895 (100.0)	5.6	1190 (84.2)	201 (14.2)	23 (1.6)

In last four to five years, the local population have realised the value of fish farming. This demonstration effort has come from a few private entrepreneurs who are in prawn culture initiated by OMCADA. In small ditches and ponds, they have put prawn and other fishes and are getting a good catch. The local population has realised the economic gains of the prawn too, although for them it is not possible to change the water or to bring sea water.

The organised ponds get two crops of prawn in a year — March to August and September to February. The prawns are sold to company agents in Jagatjore market. The prices are fixed according to the quality of the fish. The big size tiger prawns of 6-8 inches, caught from ponds and creeks are sold at a rate of Rs. 150 per kg. The useful catch from the prawn farm are 3-4 inches large and the price is Rs. 100 per kg. The prawns are sold after removing the head and are known as "headless" prawns. At Jagatjore, fish are collected between 4 pm to 8 pm and are taken to Jamboo by motorized boats from where they are transported to Paradip either in galvanised trucks or in motor launch. In Jamboo, proper packing is done with ice. In Paradip port, the international vessels come to carry the fish to Taiwan, Korea, Japan and Holland. The major fish companies operating in this area are Flacon and Golden.

Fish other than prawns are sold in different ways. Good quality fish — Scampy, Chabra and Monodon are sold to the fish agents who take them to nearby towns. The rate of the fish of more than 6 inches in length varies from Rs. 10 to 20 while the small fishes are sold by the fisher women in the surrounding villages at the rate of Rs. 6-7 per kg.

The payment is made by the fish collecting agencies on a weekly basis and the account is maintained properly. So far, no major complaint has been reported about the settlement of account. The fish companies also provide interest free advance money to the Fishermen Co-operatives and to individual fishermen. The amount is between Rs. 500 to Rs. 5000, depending upon the repayment capacity of the fisherman, given at any time during the year and recovered in form of fish during the season. The fishermen are obliged to sell their fish to agents who advance loan. The price paid in this situation are 20 to 25 per cent less than the usual market price.

Table 6.2 : Percentage Distribution of Respondents' Occupation by Annual Income

(Site : Jagatjore)

	Annual Income			
Occupation	*< 3000*	*3000-5000*	*Above 5000*	*Total*
Fishing	6	2	4	12
	50.00	16.67	33.33	100.00
	25.00	7.41	17.39	16.22
Cultivation	11	20	8	39
	28.21	51.28	20.51	100.00
	45.83	74.07	34.78	52.70
Other Occupation	7	5	11	23
	30.43	21.74	47.83	100.00
	29.17	18.52	47.83	31.08
Total	24	27	23	74
	32.43	36.49	31.08	100.00
	100.00	100.00	100.00	100.00

Analysis of Individual Responses

General economic condition of the villagers of this area is not very sound. Analysis of the sample data collected from 74 representative households (almost five per cent of the total households in these villages) in Table 6.2 indicates that 43 per cent of them have total annual income less than

Rs. 3000 which is much below the poverty line. Only 19 per cent of the households have their annual income between Rs. 3000 to Rs. 5000. Those who are above poverty line and who are near to the poverty line are the Oriya population from Gogua, Bandhapada, Kharianta, Nanjura and Sugal villages. Those below the annual income of Rs. 3000 are mostly the Bengali fisherman, Santhals and Schedule Castes.

More than half of the sampled households occupation is agriculture whereas almost one-third of them depend upon labour job they have to take up in fields of their fellow villagers as well as they go out of the village to work. Only 12 per cent households have fishing as primary occupation, although those who fish also earn their part of earning from labour wage (see Table 6.3). Santhals have indicated their primary occupation as fishing. Fishing being the seasonal occupation the respondents do not consider it a primary one. Those in agriculture belong mostly to other caste category although labourers are from all the caste categories. They don't own land and three-fourth of them live in thatched huts and only one in the sample live in *pucca* house with plastered *pucca* wall and tiled roof (see Table 6.4).

The caste group-wise education of the respondents (Table 6.5) shows that 33.8 per cent of the respondents are illiterate and 50.0 per cent of them have education below primary level. Only 1.4 per cent are educated beyond secondary level. Educationally none of the fisherman and the Scheduled Castes/Scheduled Tribes have beyond secondary level.

Table 6.6 gives a clear idea that little more than 70 per cent of the respondents are of the view that women will be contributing a great deal in the activities related to shrimp farming. They accept women's role in guarding the farm, collecting the seeds, mixing the feed and providing feed to the pond.

Of the total respondents, 93.2 per cent want to handle the pond independently and do not want to work on fixed wage whereas 6.8 per cent of them offer themselves to work on fixed wage. All of the fisherman and Scheduled Castes/Scheduled Tribes prefer to have independent handling of pond while 10.2 per cent of other castes did not want to take risk of loosing money and wanted to have fixed income as wage (see Table 6.7).

Although 64.9 per cent of the respondents are confident that they can take care of their pond by themselves, 35.1 per cent required assistance from others (see Table 6.8). The fisherman and other caste respondents have higher degree of confidence (72.7 per cent and 65.3 per cent respectively), only 57.1 per cent of the Scheduled Castes/Scheduled Tribes respondents indicated that they can take care of a pond themselves. But all of them required help in terms of land and money required for the shrimp culture, the technical support in terms of facilities to change the brackish

Table 6.3 : Percentage Distribution of Respondents' Caste Category by Occupation

(Site : Jagatjore)

Caste Category	*Occupation*			*Total*
	Fishing	*Cultivation*	*Other*	
Fishermen	3	4	4	11
	27.27	36.36	36.36	100.00
	25.00	10.26	17.39	14.86
Scheduled Castes/ Scheduled Tribes	4	2	8	14
	28.57	14.29	57.14	100.00
	33.33	5.13	34.78	18.92
Other Castes	5	33	11	49
	10.20	67.35	22.45	100.00
	41.67	84.62	47.83	66.22
Total	12	39	23	74
	16.22	52.70	31.08	100.00
	100.00	100.00	100.00	100.00

Table 6.4 : Percentage Distribution of Respondents' Caste Category by House-Type

(Site : Jagatjore)

Caste Category	*House-Type*			*Total*
	Thatched Hut	*Kuchcha*	*Pucca Having Thatched Roof/ Pucca*	
Fishermen	10		1	11
	90.91		9.09	100.00
	17.86		33.33	14.86
Scheduled Castes/ Scheduled Tribes	14			14
	100.00			100.00
	25.00			18.92
Other Castes	32	15	2	49
	65.31	30.61	4.08	100.00
	57.14	100.00	66.67	66.22
Total	56	15	3	74
	75.68	20.27	4.05	100.00
	100.00	100.00	100.00	100.00

Table 6.5 : Percentage Distribution of Respondents' Caste Category by Education

(Site : Jagatjore)

Caste Category	*Illiterate*	*Below Primary Level*	*Below Secondary Level*	*Beyond Secondary Level*	*Total*
	Education				
Fishermen	4 36.36 16.00	6 54.55 16.22	1 9.09 9.09		11 100.00 14.86
Scheduled Castes/ Scheduled Tribes	7 50.00 28.00	7 50.00 18.92			14 100.00 18.92
Other Castes	14 28.57 56.00	24 48.98 64.86	10 20.41 90.91	1 2.04 100.00	49 100.00 66.22
Total	25 33.78 100.00	37 50.00 100.00	11 14.86 100.00	1 1.35 100.00	74 100.00 100.00

Table 6.6 : Percentage Distribution of Respondents' Caste Category by Women Help

(Site : Jagatjore)

Caste Category	*Feeding*	*Farm Guarding*	*Marketing*	*No Help-House Wife Only*	*No Response*
	Women Help				
Fishermen	8 21.62	5 16.13	2 14.29	1 4.76	1 25.00
Scheduled Castes/ Scheduled Tribes	6 16.22	6 19.35	3 21.43	4 19.05	1 25.00
Other Castes	23 62.16	20 64.52	9 64.29	16 76.19	2 50.00
Total	37 100.00	31 100.00	14 100.00	21 100.00	4 100.00

'N' is more because of multiple choice.

water and measuring the salinity regularly, the management for common supply, maintenance and marketing of the product. They to require technical training for the proper way of doing prawn farming — how to select quality seed, proper way of feed, the implications of salinity and about the disease and precaution to be taken in shrimp farming.

The respondents have their own views about running the co-operative of those owning the ponds. 74.3 per cent of the respondents want to run the co-operatives with the help of the government agencies, mainly the Department of Fisheries, only one-fourth of them (25.7 per cent) did not want any government help (including the fisheries department) in co-operative affairs (see Table 6.9). However, the analysis given in Table 6.10 clearly indicates that while all the respondents are inclined to form co-operatives of pond owners, 23 per cent insisted to form co-operative of their own caste(s). This notion is strong among Scheduled Castes/ Scheduled Tribes as compared to other castes.

Co-operatives and Co-operation

The word co-operative is not alien to many villagers who have taken loan from Lanjura Co-operative Society and Belakanda Service Co-operative

Table 6.7 : Percentage Distribution of Respondents' Caste Category by Attitude Towards Handling of Pond

(Site : Jagatjore)

Caste Category	*Pond Handling*			*Total*
	Independently	*On Fixed Wages*	*On Co-operative Basis*	
Fishermen	11			11
	100.00			100.00
	15.94			14.86
Scheduled Castes/ Scheduled Tribes	14			14
	100.00			100.00
	20.29			18.92
Other Castes	44	5		49
	89.80	10.20		100.00
	63.77	100.00		66.22
Total	69	5		74
	93.24	6.76		100.00
	100.00	100.00		100.00

Table 6.8 : Percentage Distribution of Respondents' Caste Category by Attitude Towards Pond Care

(Site : Jagatjore)

Caste Category	*Pond Care*		*Total*
	Can Take Himself/ Herself	*Assistance Required*	
Fisnermen	8	3	11
	72.73	27.27	100.00
	16.67	11.54	14.86
Scheduled Castes/ Scheduled Tribes	8	6	14
	57.14	42.86	100.00
	16.67	23.08	18.92
Other Castes	32	17	49
	65.31	34.69	100.00
	66.67	65.38	66.22
Total	48	26	74
	64.86	35.14	100.00
	100.00	100.00	100.00

Table 6.9 : Percentage Distribution of Respondents' Caste Category by Attitude Towards Running of Co-operative Society

(Site: Jatagjore)

Caste Category	*Running of Co-operative*			*Total*
	With Government / Fish. Dept. Help	*Without Government Help*	*No Response*	
Fishermen	10	1		11
	90.91	9.09		100.00
	18.18	5.26		14.86
Scheduled Castes/ Scheduled Tribes	9	5		14
	64.29	35.71		100.00
	16.36	26.32		18.92
Other Castes	36	13		49
	73.47	26.53		100.00
	65.45	68.42		66.22
Total	55	19		74
	74.32	25.68		100.00
	100.00	100.00		100.00

Table 6.10 : Percentage Distribution of Respondents' Caste Category by Joining of Co-operatives with Own Caste/Other Caste Members

(Site : Jagatjore)

Caste Category	*Joining of Co-operatives* With Own Caste Members	With other Caste Members	No Response	*Total*
Fishermen	3	8		11
	27.27	72.73		100.00
	17.65	14.04		14.86
Scheduled Castes/	5	9		14
Scheduled Tribes	35.71	64.29		100.00
	29.41	15.79		18.92
Other Castes	9	40		49
	18.37	81.63		100.00
	52.94	70.18		66.22
Total	17	57		74
	22.97	77.03		100.00
	100.00	100.00		100.00

Society. There are 930 members of Belakanda Service Co-operative Society alone. The Society provides crop loan against mortgage of land. The money is repaid after the harvest with 10 per cent annual interest on it. But their experiences are different.

There are two co-operative societies — one registered and other unregistered. The registered society is registered with District Fisheries Officer who also acts as the Registrar for fisheries co-operative societies.

Maa Ranachandi Primary Prawn Culture and Marketing Co-operative Society with 22 members from 10 villages of this area was formed on the initiative of the Fisheries Extension Officer (FEO), Mahakalpada Block. The FEO is the President and Chief Executive Officer, Brackish Water Fisheries Development Agency (BFDA) is one of the six members of the governing body. A local member is the secretary of the society.

Another unregistered society **Sidheswari Co-operative Society** with its head office in Gogua village (details are given in the Appendix-6) is operating on an unauthorised land.

It has been reported that the members of both the societies belong to different villages and different castes but have come together to form a society. Sidheswari Co-operative Society being a large society has appointed a care taker on share basis. The members go to catch the fish together.

At Jagatjore site, together effort of forming a co-operative society for prawn culture on the acquired land they had taken on lease from revenue department was made 5 years ago for paddy cultivation. **Nanjura Sambaya Samiti** is Nanjura based registered Agricultural Credit Co-operative Society established more than 10 years ago and has about 1200 members. Although this society is said to function smoothly, presently it has been taken over because of some problems during the last election of the society.

For most of the villagers, co-operative means getting enrolled for the membership and asking for agricultural loan.

The sense of working together and sharing the benefit accordingly is somewhat missing. They do not have such notion because they have not experienced such a co-operation. The Gogua Fishermen Society and the Co-operative activity undertaken by the Sarpanch with seven other members from outside this area has different experiences for the members as well as others who know about it. The Co-operative run by the Sarpanch is the effort of seven rich people (friends and relatives from Cuttack and Kendrapada) who have put their financial resources together to take 12 ponds of 60 hectare of land and employed labourers to get benefit. Out of these only the local member takes proper care and rest are there to share the benfit. The Fishermen Co-operative Society has many uninterested, a few genuinely interested, and very few who really work for the pond. Sharing of benefit is equal but not the sharing of responsibilities. It is not a happy affair. The villagers, by and large, have a feeling that any co-operative effort will have the same fate, if not in the beginning after a few years.

BRACKISH WATER SITE : PANASPADA

Panaspada forms part of Krushna Prasad Block of Puri District (see Maps 6.4 and 6.5). Nineteen islands of various sizes in the wellknown Chilka Lake constitute this Block's landmass; only a small strech towards the east having a land border and it is at this stretch that Panaspada brackish water site is located. The site is about 10 km from Brahmagiri Police Station and 40 km from Puri city by road. The other approach to site from Baligaon in the north is by waterways. The Block Headquarters about eight km from Panaspada brackish water site and is approachable only through a combined network of road and waterways. From site to Satapada, one has to use the road and then pass the creek by regular motor boat service (available four times a day and then one has to walk for three km for

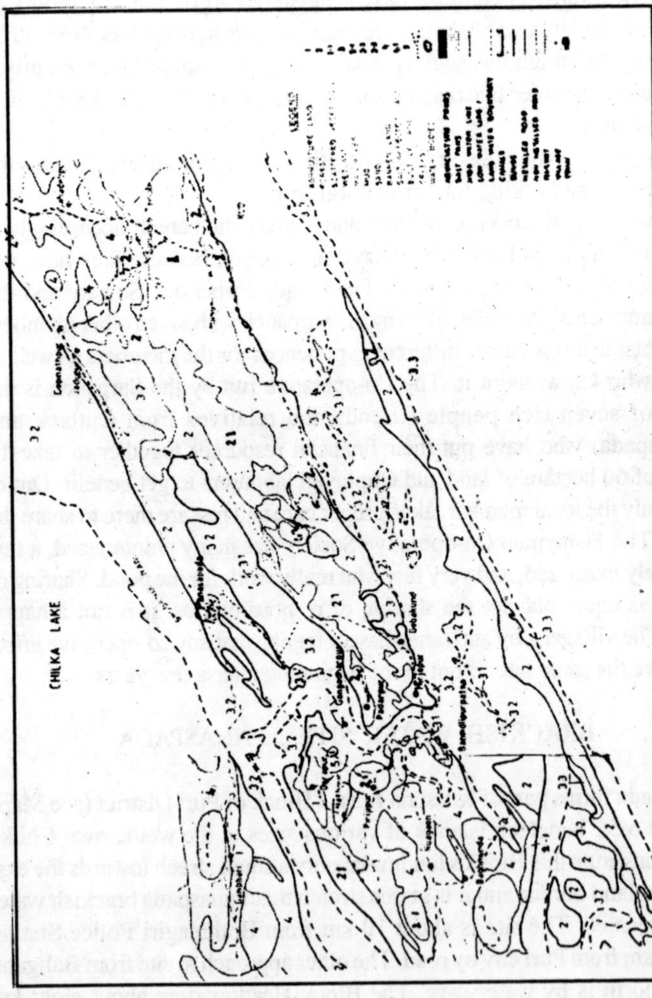

Map 6.4 : Land Use Map Site : Panaspada

Source : Remote Sensing Applications Group, ISRO, Ahmedabad, 1990.

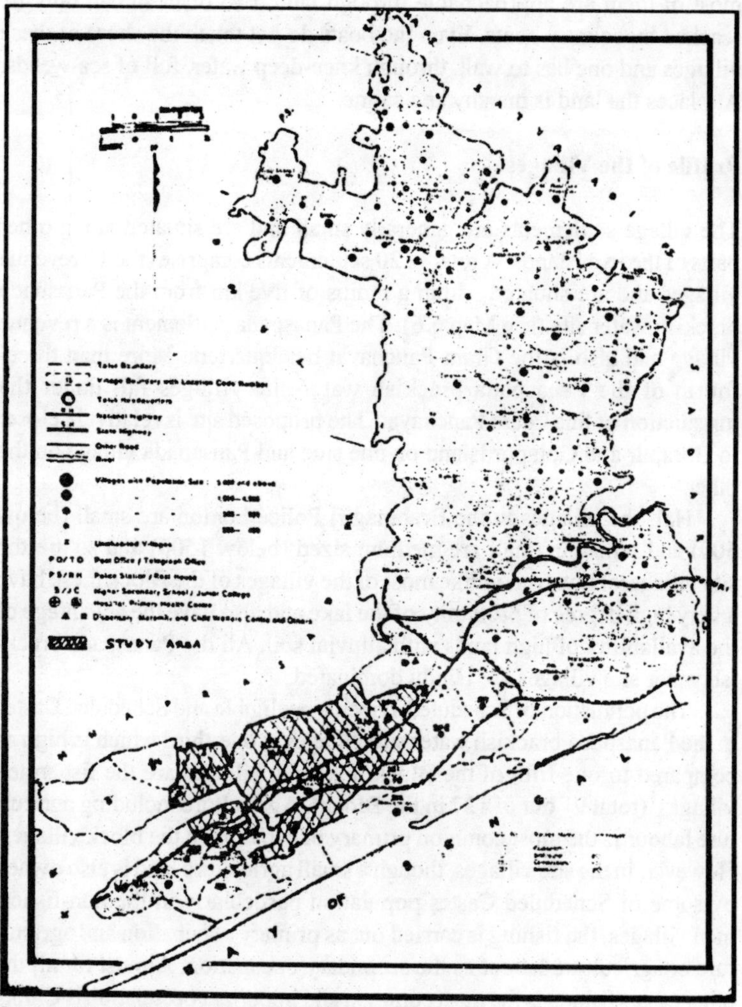

Based upon Survey of India Map with the permission of the Surveyor General of India.
The Territorial States of India extended into the Sea to a distance of twelve nautical miles measured from the appropriate base line

Source : Census of India, 1981.

Map 6.5 : Brahmagiri P.S. District Puri, Orissa
Site : Panaspada

reaching the Block Headquarters. A total 10 revenue villages are located within a radius of five km from the Panaspada brackish water site. While most of them are approachable through land, two of them can only be reached by country boats. Even the boats do not touch the shore of these villages and one has to walk through knee-deep water, full of sea-weeds. At places the land is marshy and saline.

Profile of the Villages

The village settlements are generally small and are situated along other parts of the two islands. A total of 20 settlements comprises the 10 revenue villages and are situated within a radius of five km from the Panaspada brackish water site (see Map 6.6). The Panaspada settlement is a revenue village and also is the Gram Panchayat headquarters. More than three-fourth of the Panaspada brackish water site villages fall under the jurisdiction of this Gram Panchayat. The proposed site is relatively closer to Gorapur and Golapur island on one side and Pansapada village on the other.

Half the villages in the Brahmagiri Police Station are small (below 500) and remaining half are medium sized (below 1500) and so are the site villages. However, unlike most of the villages of the Block these have a very high degree of proximity of the lake and also have the advantage of the availability of high lands with alluvial soil. All the Panaspada brackish water site villages are Hindu dominated.

The population of Scheduled Tribes is negligible and Scheduled Castes in the Panaspada brackish water site villages is one-third which is high as compared to one-fifth of the Block. Five site villages are the fishermen villages (total 91 out of 127 in the Block). Agriculture including agriculture labour is the most common primary occupation in the block villages. However, in the site villages, though a small agriculture land is also owned by some of Scheduled Castes population particularly in the non-fishermen villages, the fishing is carried out as primary occupation and agriculture or agriculture labour as the secondary occupation. Almost for all the other castes fishing is the most common and important occupation is Chilka Lake.

About two dozen households from among the Panaspada brackish water site villages cultivate prawn. They do so either in their own ponds with the support under IRDP/ERRP schemes or in Gram Panchayat's leased brackish water ponds. Some households even go for purchase of prawn seed from Paradip and Gopalpur and sell it to the fellow fish farmers. The

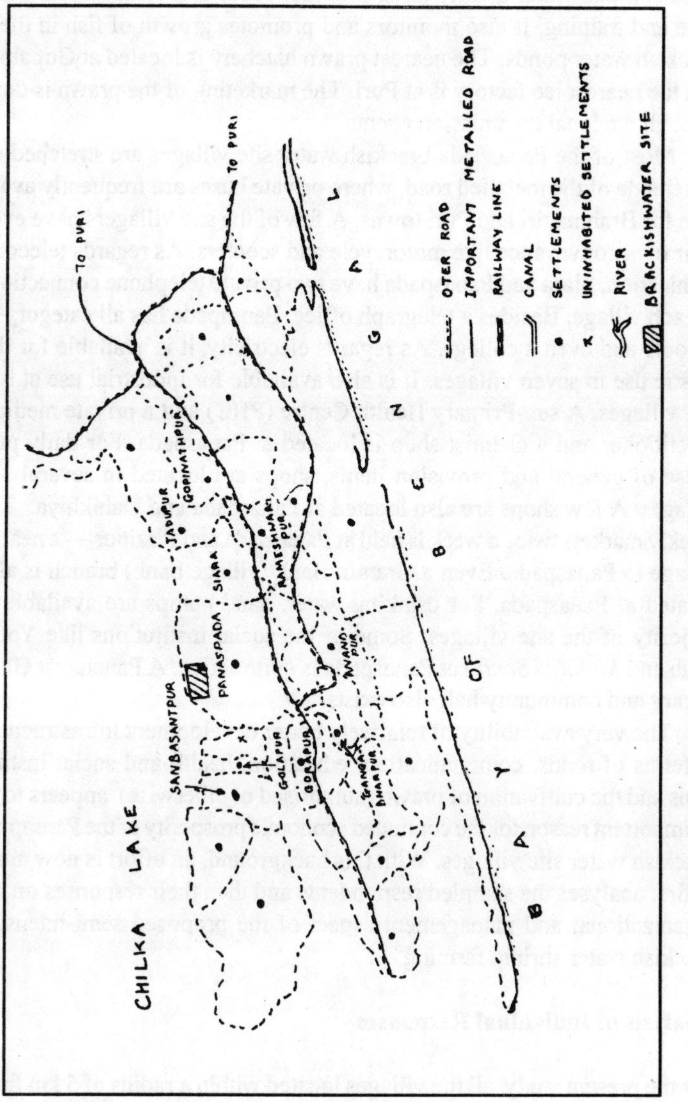

Map 6.6 : Map showing villages visited at Panaspada Brackish Water Site

Brackish Water Fisheries Development Agency's Sunamni office undertakes soil and water quality tests, provides seeds and all technical guidance and training. It also monitors and promotes growth of fish in these brackish water ponds. The nearest prawn hatchery is located at Gopalpur and the nearest ice factory is at Puri. The marketing of the prawn is done through the local commission agents.

Most of the Panaspada brackish water site villages are stretched on either side of the metalled road, where private buses are frequently available for Brahmagiri and Puri towns. A few of the site villagers have even their own conveyance like motorcycle and scooters. As regards telecommunication, Siara and Panaspada have two private telephone connections in each village. Besides a telegraph office, Panaspada has all category of schools and even a college. As regards electricity, it is available for domestic use in seven villages. It is also available for industrial use at two site villages. A sub-Primary Health Centre (PHC) and a private medical practitioner and a chemist shop is located at Panaspada. For daily purchase of general and provision items, shops are located in several site villages. A few shops are also located at Panaspada and Dahikhiya. The weekly markets twice a week is held at Panaspada and Pirzipur — a nearby village to Panaspada. Even a Gramin Bank (village bank) branch is also located at Panaspada. For drinking water, hand pumps are available in majority of the site villages. Some of the social institutions like Youth Club and *Mahilya Samiti* at Panaspada is quite active. A Panchayat Ghar, library and community hall also exists.

The very availability of relatively better development infrastructure, in terms of roads, communication, education, health and social institutions and the cultivation of prawn (authorised or otherwise), appears to be an important reason for the continued economic prosperity of the Pansapada brackish water site villages. With this background, an effort is now made to first analyses the sampled respondents and then their responses on the organizational and management aspect of the proposed semi-intensive brackish water shrimp farming.

Analysis of Individual Responses

For the present study, all the villages located within a radius of 5 km from the site were selected. From each village, six respondents were selected for the household level survey. For the Panaspada site, therefore, we had a total of 60 respondents. Before, any inferences are drawn for the study, it is necessary to know the extent to which the sampled respondents represent the potential beneficiaries of the project.

Table 6.11 reveals that there are a total of 941 households in the ten villages selected for the study. Around one-third of these households are of Fisherman; one-fifth (21.68 per cent) of SC/ST; and the remaining little less than half (46.55 per cent) are of other castes. The average household size is 6.6. The Socio-Economic Status (SES) of 57.81 per cent households is on lower side; 38.68 per cent of middle level and; remaining 3.51 per cent on upper side. Further analysis of this table indicates that large number of Fisherman and SC/ST households enjoy SES level, which is substantially high as compared to other castes house- holds. This is further a point of clarification that fishing is one of the most important paying occupation undertaken by almost all households irrespective of their caste affiliations. With this in mind, let us now analyse the characteristics of the sampled respondents, who have been choosen from the villages.

Table 6.15 indicates that education-wise 50 per cent of the respondents are illiterate; 35 per cent below primary and; the remaining of below secondary level. It was elsewhere observed that none of the respondents had a technical qualification. Also elsewhere observed was that all of them are Hindus. The table under reference also indicate that castewise 15 per cent of them are fishermen; 45 per cent are SC/ST; and only 40 per cent are from other castes.

Table 6.12 indicates that in income-wise 36.67 per cent of the respondents annually earn below Rs. 3000; 48.33 per cent below Rs. 5000 and only 15 per cent above Rs. 5000. 58.33 per cent of this income is from fishering; 26.67 per cent from other occupation; and only 15 per cent comes from cultivation. The fact emerging from this analysis is that fishing is the most paying occupation in the study area. Another Table 6.13 highlights that fishing is a very important occupation among both fishermen and SC/ST. Among other castes, cultivation is not that very important an occupation.

Table 6.14 indicate that 90 per cent of the respondents stay in thatched hut, and 33 per cent in *kuchcha* house; and only 1.67 per cent is *pucca* having thatched roof. The analysis of this table also indicates that fishermen stay in relatively better *kuchcha* houses.

Little less than two-third of the respondents have no land; and the remaining one-third have land below 3 acres (Table 3.3). All these lands are unirrigated. In spite of probing, majority of the respondents did not reveal the waterbody area possessed by them in Chilka Lake. However, none of them have a private pond (Table 6.15). Only a few respondents keep bovine like duck and poultry (Table 3.4). Table 2.2 indicates that slightly less than half of the respondents are female. Age-wise, more than half of the respondents are in the younger age group (up to 40) and the remaining in the elderly group (Table 6.21). None of the respondents indicated having taken any kind of government loan (Table 3.7). All the above, together with analysis of Table 6.11 goes to indicate that since this area is economically better off, the sampled respondents generally represent the little better off poor group of the targeted population.

It is quite interesting to find that people are generally aware of the method and system of fisheries co-operative society, one-third of the respondents were the members of such village level Primary Fishermen Co-operative Societies (PFCs) (Table 6.22). Half of the respondents have preferred to join a co-operative society having their own caste members. Other less than half have no hesitation in joining a co-operative society having even other caste members (Table 6.20). Amazingly, all of them have favoured a society having members from within their village or at the most nearby Gram Panchayat villages. Two-third of the respondents have felt that they can take care of their pond with the help of their families themselves, the remaining one-third might need some help from other progressive fishermen's families (Table 6.8). Almost all the respondents have felt that without the active help of a fisheries official, they would not be able to successfully run the co-operative society. Their major help will be required in keeping accounts and marketing and recovery of loan on the spot (Table 6.9).

A large majority of the respondents, however, have felt that the pond ownership should be individual based (Table 6.7). For growing fish, the individuals would make their own independent efforts, but for procuring seed, feed and change of water and the sale of fish, the co-operative society should take the responsibility under overall supervision and control of fisheries officials. The fisheries officials have been considered to be the most trusted person for handling the co-operative society's money (Table 6.23). Two-third of the respondents have felt that people are aware about the high market value of prawn, and since non-fishing families have also been cultivating fish in Chilka, given the opportunity they would also jump into this activity (Table 6.24). They would like to be given training in identfying the types of diseases in prawn, conducting of soil and water quality tests. They would also need monetary help *i.e.* loan for excavation of the ponds. As regards women's participation, their help will be required for fish feeding and supervision of the Fish Farms (Table 6.16).

Co-operatives and Co-operation

Out of 10 study villages, it is found that there are five Primary Fishermen Co-operative Societies (PFCs) in operation in Chilka Lake for the past two to eight years. Budhima Primary Fishermen Co-operative Scoiety, Gangadharpur is a registered society with Central Fisherman Co-operative Society (CFC), Balugaon. The CFC has all the legal ownership rights of the entire *Shairat* of Chilka Lake area. CFC this year has renewed the lease of 500 acres of Budhima PFC after charging Rs. 20,000 as lease value. The Society has to pay 10 per cent as the annual incremental lease value. The Society was established in 1982-83 by Late Ling Raj, one of the residents of this village who was educated upto intermediate or 12th grade and was in his early twenties. Lakhmidhar of this village in his forties, is the present secretary. The Society has all the 105 members belonging to fishermen and Scheduled Castes of this village. The President of the Society also belongs to this village. It is pertinent to know that the Society has been successfully operating since its inception therefore, is important to understand its methods and modes of operation.

Generally, the huge amount for paying the lease value to CFC is taken as loan by the Secretary and President from a local commission agent. The unwritten understanding is that they pay back 25 per cent extra on the amount taken and also once a week, the fish catch will have to be sold through this commission agent. The fish catch is undertaken every day. Country boats owned by three members is also used for fishing. While those who use their boat would get 25 per cent extra as their share and others would get on an equal basis. If a member does not go for fishing on any day, he foregoes that day's shares. On all days of the week, the members are free to sell their fish catch to anybody except on one day of the week, when it is given to the commission agent in lieu of the advance taken from him. In case anybody does not come for fishing on that day, he is charged for labour and this amount is deducted from his share. The President, Secretary and members of this Society feel that their income is decreasing day by day now as they do not get adequate fish catch these days from the lake. The Society in fact does not cultivate the fish, whatever in natural course is the fish growth, they just catch that only. No efforts have been made for fish feeding in the waterbody area allotted to them.

The other three Primary Fishermen Co-operative Societies, which are yet to be registered, are located in the study villages. Laxmi Nursingh

Primary Fishermen Co-operative Society, Gopinathpur is a large and recently formed society, while the other two have been working for the past two to six years. Sanbasantapur Society is not recognised becuase it consists of members from non-fishing castes also.

Some members of all these societies have had some training in fishing a few years ago. However, they have been practicing it since childhood which they feel to be the best training. These three societies are though not registered and do not pay the lease money to CFC, but their mode of understanding method of operation is almost the same as followed by the Budhima PFC.

Bagdevi Primary Fisherman Co-operative Society, Siara-Gola was formed in 1976. It has 108 members own. The society purchase fish and sell it to the companies at higher rates in the market. Many villagers have purchased boats, nets with the help of the local bank. These boats are used for fishing and in each boat more than one person goes. However, it is pertinent to note that the fish catch is equally divided. All this shows the feeling of co-operation among the villagers.

From Sunamni, to Satapada, there are about 52 unauthorised encroched *Gheris* (waterbody) at Chilka Lake. All such Gheries with an area of 10 to 80 ha. are managed by various groups of 5 to 10 families. Most of such Gheries are possessed by Khandait — other castes group. However, 2-3 Gheries are also exclusively possessed and managed by Scheduled Caste group. The practice of fish cultivation followed is traditional, *i.e.*, simply putting seed and after 3-4 months harvesting it. Various groups of people have been active in operation. These groups through their selected members also manage some advance money from the local traders for minor repairs of their Gheries, go to Paradip for purchase of seeds or buy seed from Fisheries Department. There does not seem to occur any major conflicts between these groups on fish catch, distribution, repayment of loan or any other matter. Mode, method and understanding of operation is the same for the Gram Panchayat tanks, which are authorised and taken on lease basis by similar groups of people.

Almost on similar lines of the proposed World Bank Project, there are already 147 brackish water ponds of 0.2 ha. size given to the poorest of the poor on an individual basis under ERRP (100 per cent subsidy) scheme of the State situated at Khamdulpur, a village at a distance of eight km from Panaspada. More, so, under IRDP, there are other 173 such ponds. Again, most of them in clusters are situated in the villages of Jadupur, Panaspada, Gorapur and three other nearby site villages. The subsidy component is now further relaxed and the benefits of the scheme is now also available to all private pond owners irrespective of their income.

Whatever is the subsidy component, one point is very clear that the poor of this area are aware about prawn cultivation. Further, it is important to know that in spite of free allotment of 147 properly constructed brackish water ponds, these poor pond owners do not seem to have gained anything substantial from these ponds. Their economic conditions remain unchanged. The Sunamni Fisheries Office is quite close to these clustered ponds, cultivators get all technical like soil and water tests) help from them. This is primarily because of the fact that these pond owners simply put some seeds and harvest after 3-4 months. For want of money, proper control and management system, farmers are not able to cultivate prawn in a scientific manner. Further a point to keep in view is that fish farmers stay in their permanent settlements and come for fishing and then go back to their homes. This trend is universal for all the co-operative societies, larger Gheries of Chilka and leased in Gheries of Gram Panchayats.

Table 6.11 : Distribution of Caste Category by Households, Population and Socio-Economic Status in Study Villages

(Site : Panaspada)

Caste Category	*Number of Household*	*Total Population*	*Household Size*	*Socio Economic Status*		
				Lower	*Middle*	*Upper*
Fishermen	299	2075	6.9	129	150	20
	100.00			43.14	50.17	6.69
	31.77	33.62		23.71	41.21	60.61
Scheduled Castes/ Scheduled Tribes	204	1188	5.8	155	39	10
	100.00			75.98	19.12	4.90
	21.68	19.25		28.49	10.71	30.30
Other Castes	438	2909	6.6	260	175	3
	100.00			59.36	39.95	0.68
	46.55	47.13		47.79	48.08	9.09
Total	941	6172	6.6	544	364	33
	100.00			57.81	38.68	3.51
	100.00	100.00		100.00	100.00	100.00

Table 6.12 : Percentage Distribution of Respondents' Occupation by Annual Income

(Site : Panaspada)

	Annual Income			
Occupation	*< 3000*	*3000-5000*	*Above 5000*	*Total*
Fishing	15 42.86 68.18	17 48.57 58.62	3 8.57 33.33	35 100.00 58.33
Cultivation	6 66.67 27.27	3 33.33 10.34		9 100.00 15.00
Other Occupation	1 6.25 4.55	9 56.25 31.03	6 37.50 66.67	16 100.00 26.67
Total	22 36.67 100.00	29 48.33 100.00	9 15.00 100.00	60 100.00 100.00

Table 6.13 : Percentage Distribution of Respondents' Caste Category by Occupation

(Site : Panaspada)

	Occupation			
Caste Category	*Fishing*	*Cultivation*	*Other*	*Total*
Fishermen	7 77.78 20.00	1 11.11 11.11	1 11.11 6.25	9 100.00 15.00
Scheduled Castes/ Scheduled Tribes	23 85.19 65.71	1 3.70 11.11	3 11.11 18.75	27 100.00 45.00
Other Castes	5 20.83 14.29	7 29.17 77.78	12 50.00 75.00	24 100.00 40.00
Total	35 58.33 100.00	9 15.00 100.00	16 26.67 100.00	60 100.00 100.00

Table 6.14 : Percentage Distribution of Respondents' Caste Category by House Type

(Site : Panaspada)

Caste Category	*House-Type*			*Total*
	Thatched Hut	*Kuchcha*	*Pucca Having Thatched Roof/ Pucca*	
Fishermen	7	2		9
	77.78	22.22		100.00
	12.96	40.00		15.00
Scheduled Castes/ Scheduled Tribes	26	1		27
	96.30	3.70		100.00
	48.15	20.00		45.00
Other Castes	21	2	1	24
	87.50	8.33	4.17	100.00
	38.89	40.00	100.00	40.00
Total	54	5	1	60
	90.00	8.33	1.67	100.00
	100.00	100.00	100.00	100.00

Table 6.15 : Percentage Distribution of Respondents' Caste Category by Education

(Site : Panaspada)

Caste Category	*Education*				*Total*
	Illiterate	*Below Primary Level*	*Below Secondary Level*	*Beyond Secondary Level*	
Fishermen	7	2			9
	77.78	22.22			100.00
	23.33	9.52			15.00
Scheduled Castes/ Scheduled Tribes	15	7	4	1	27
	55.56	25.93	14.81	3.70	100.00
	50.00	33.33	57.14	50.00	45.00
Other Castes	8	12	3	1	24
	33.33	50.00	12.50	4.17	100.00
	26.67	57.14	42.86	50.00	40.00
Total	30	21	7	2	60
	50.00	35.00	11.67	3.33	100.00
	100.00	100.00	100.00	100.00	100.00

Table 6.16 : Percentage Distribution of Respondents' Caste Category by Women Help

(Site : Panaspada)

Caste Category	*Women Help*				
	Fish Feeding	*Farm Guarding*	*Marketing*	*No Help-House Wife Only*	*No Response*
Fishermen	2 8.00	3 12.00		6 26.09	1 100.00
Scheduled Castes/ Scheduled Tribes	11 44.00	13 52.00	3 60.00	8 34.78	
Other Castes	12 48.00	9 36.00	2 40.00	9 39.13	
Total	25 100.00	25 100.00	5 100.00	23 100.00	1 100.00

'N' is more because of multiple choice.

Table 6.17 : Percentage Distribution of Respondents' Caste Category by Attitude Towards Handling of Pond

(Site : Panaspada)

Caste Category	*Pond Handling*			
	Independently	*On Fixed Wages*	*On Co-operative Basis*	*Total*
Fishermen	8 88.89 13.79	1 11.11 100.00		9 100.00 15.00
Scheduled Castes/ Scheduled Tribes	26 96.30 44.83		1 3.70 100.00	27 100.00 45.00
Other Castes	24 100.00 41.38			24 100.00 40.00
Total	58 96.67 100.00	1 1.67 100.00	1 1.67 100.00	60 100.00 100.00

Table 6.18 : Percentage Distribution of Respondents' Caste Category by Attitude Towards Pond Care

(Site : Panaspada)

Caste Category	*Pond Care*		*Total*
	Can Take Himself/ Herself	*Assistance Required*	
Fishermen	7 77.78 16.67	2 22.22 11.11	9 100.00 15.00
Scheduled Castes/ Scheduled Tribes	16 59.26 38.10	11 40.74 61.11	27 100.00 45.00
Other Castes	19 79.17 45.24	5 20.83 27.78	24 100.00 40.00
Total	42 70.00 100.00	18 30.00 100.00	60 100.00 100.00

Table 6.19 : Percentage Distribution of Respondents' Caste Category by Attitude Towards Running of Co-operative Society

(Site : Panaspada)

Caste Category	*Running of Co-operative*			*Total*
	With Government / Fish. Dept. Help	*Without Government Help*	*No Response*	
Fishermen	9 100.00 15.00			9 100.00 15.00
Scheduled Castes/ Scheduled Tribes	27 100.00 45.00			27 100.00 45.00
Other Castes	24 100.00 40.00			24 100.00 40.00
Total	60 100.00 100.00			60 100.00 100.00

Table 6.20 : Percentage Distribution of Respondents' Caste Category by Joining of Co-operatives with Own Caste/Other Caste Members

(Site : Panaspada)

Caste Category	*Joining of Co-operatives* With Own Caste Members	*With other Caste Members*	*No Response*	*Total*
Fishermen	7 77.78 21.88	1 11.11 4.17	1 11.11 25.00	9 100.00 15.00
Scheduled Castes/ Scheduled Tribes	21 77.78 65.63	5 18.52 20.83	1 3.70 25.00	27 100.00 45.00
Other Castes	4 16.67 12.50	18 75.00 75.00	2 8.33 50.00	24 100.00 40.00
Total	32 53.33 100.00	24 40.00 100.00	4 6.67 100.00	60 100.00 100.00

Table 6.21 : Distribution of Respondents by Age

(Site: Panaspada)

Age Group	*Frequency (N)*	*Percentage (%)*
Below 20	2	3.33
20-40	31	51.67
Above 40	27	45.00
Total	60	100.00

Table 6.22 : Distribution of Respondents by Present Co-operative Membership

(Site : Panaspada)

Member of Co-operative	*Frequency (N)*	*Percentage (%)*
Yes	19	31.67
No	41	68.33
Total	60	100.00

Table 6.23 : Distribution of Respondents by People Trusted

(Site : Panaspada)

People Trusted	*Frequency (N)*	*Percentage (%)*
Fisheries/Govt. Officials	46	76.67
Elected Members	2	3.33
Traditional Caste Leaders	—	—
None-Self Only	3	5.00
Secretary Co-operatives	9	15.00
Total	60	100.00

Table 6.24 : Distribution of Respondents by Non-Fishing Family Participation in the Project

(Site : Panaspada)

Non-Fishing Family Like to Participate	*Frequency (N)*	*Percentage (%)*
Yes	44	73.33
No	16	26.67
Total	60	100.00

Table 6.25 : Distribution of Respondents by Pond Ownership

Site : Panaspada

Pond Ownership	*Frequency (N)*	*Percentage (%)*
No Pond	60	100.00
One Pond	—	—
Two Pond	—	—
Three Pond	—	—
Total	60	100.00

Table 6.26 : Distribution of Respondents by Type of Co-operatives Help

(Site : Panaspada)

Co-operatives Help Required	*Frequency (N)*	*Percentage (%)*
Loan	55	26.57
Technical Like Soil/Water Testing	49	23.67
Managerial	56	27.05
Extn. and Training	45	21.74
No Response	2	0.97
Total	207	100.00

'N' is more because of multiple choice.

7

ANDHRA PRADESH

In this Chapter the analysis of two Andhra Pradesh sites namely, P.T. Palem and C.G. Palem has been presented. The data all drawn from site visits, group discussions at village level, and individual interviews of the potential beneficiaries and Department of Fisheries officials. Observations were also made in the field by the research team.

BRACKISH WATER SITE : P.T. PALEM

The proposed P.T. Palem brackish water site is approximately six km from the village P.T. Palem, and it is located near Kothatummalapalem (K.P.T. Palem or new P.T. Palem), a hamlet of village Palletummalapalem (P.T. Palem) under Bandar Taluka of Krishna District in Andhra Pradesh. It is nearly 36 km away from Machilipatnam. It is located almost on the tip of a project geographical area of coast known as 'Tip of Divi'. It is a part of a large low lying area at a triangle of the Bay of Bengal and the tributary of River Krishna. There is a bund between all the villages on the eastern side and the river on the western side. Road comes up to 2 km from the site and bus service is available in fair weather. Almost eight to nine months in a year the condition of the road is such that no bus service is possible. After the May 1990 cyclone of Andhra Pradesh the road was under repair during the second week of December 1990 when the team visited these villages (see Maps 7.1 and 7.2).

Profile of the Villages

For the purpose of the present study eight villages in the radius of 5-7 km were selected as sufficient number of villages were not falling within five km radius (see Map 7.3). These villages as listed in Appendix-4 are hamlets of two revenue villages; Palletummalapalem village of Bandar Taluk and Padayadara village of Divi Taluk. All these villages are on Chinnapuram-Kammavaricheruvu road or half km to two km east to the road.

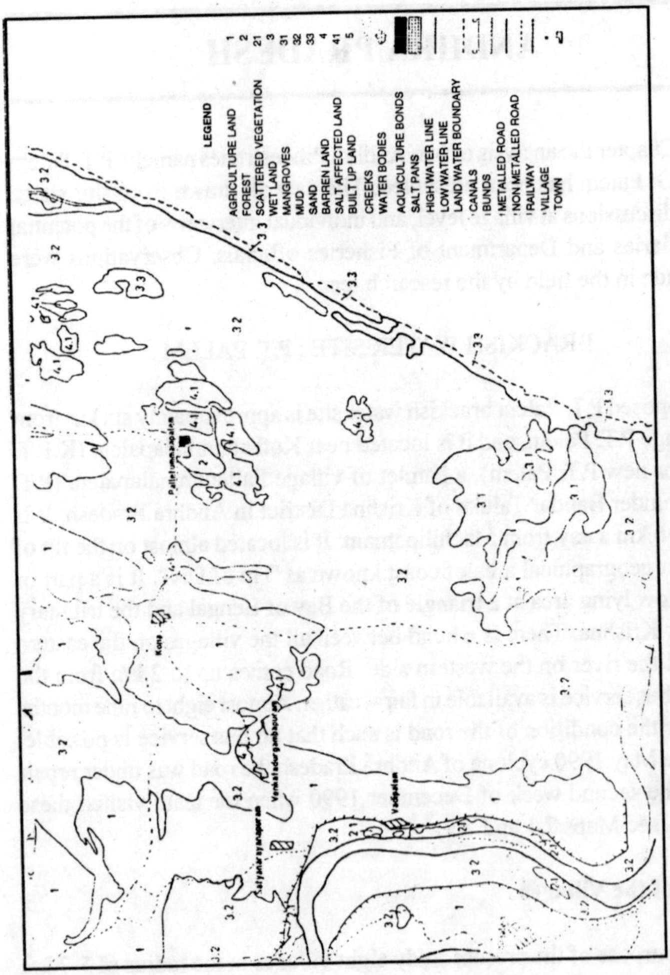

Source : Remote Sensing Applications Group, ISRO, Ahmedabad, 1990.

Map 7.1 : Landuse Map Site : P. T. Palem

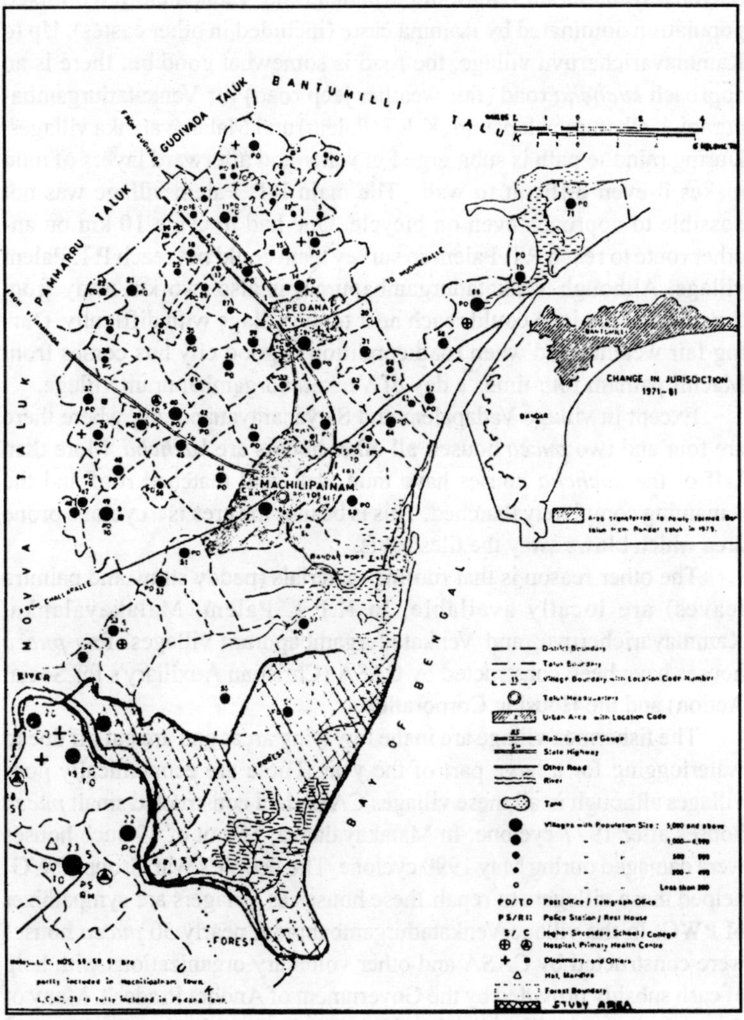

Source : Census of India, 1981.

Map 7.2 : Bandar Taluk and Part of Divi Taluk, Krishna District, Andhra Pradesh
Site : P. T. Palem

Palletummalapalem, Venkatadurgambapuram, Kothatummala-palem and Malakayalanka are the fishermen's villages. Kammavaricheruvu, Satyanarayanapuram, Bhawanipurilanka, and Vadapalem have mixed population dominated by Kamma caste (included in other castes). Up to Kammavaricheruvu village, the road is somewhat good but there is an approach *kuchcha* road (fair weather jeep road) for Venkatadurgambapuram, Kothatummalapalem (K.P.T. Palem) and Malakayalanka villages. During rain the path is submerged in water and afterward layers of mud makes it even difficult to walk. The main P.T. Palem village was not possible to approach even on bicycle. One had to cycle 10 km on another route to reach P.T. Palem so survey team could not reach P.T. Palem village. Although Venkatadurgambapuram is also two km away from the road but the jeep could reach near to the village with difficulty. During fair weather and when road condition is good city bus comes from Machilipatnam four times a day of Venkatadurgambapuram village.

Except in village Vadapalem and Satyanarayanapuram, where there are four and two *pucca* houses, all other houses are *kuchcha.* More than half of the *kuchcha* houses have mud wall, and thatched roof and the remaining completely thatched. This is beause the area is a cyclone prone area which blows away the tiles/roofs.

The other reason is that roofing materials (paddy straw and palmira leaves) are locally available. In K.P.T. Palem, Malakayalanka, Kammavaricheruvu, and Venkatadurgambapuram villages, few *pucca* houses have been constructed by CASA (Christian Auxiliary's for Social Action) and the Housing Corporation.

The fishermen village are in the low lying areas and are cut off due to waterlogging for a large part of the year. These are economically poor villages although in all these villages CASA had constructed small *pucca* houses after 1977 cyclone. In Malakayalanka, 15 out of 19 such houses were damaged during May 1990 cyclone. The People's War Group (PWG) helped these villagers to repair these houses as villagers are sympathizer of PWG. In the village Venkatadurgambapuram nearly 50 *pucca* houses were constructed by CASA and other voluntary organizations with help of cash subsidy provided by the Government of Andhra Pradesh. Many of these houses are in dilapidated condition and owners have moved out and are staying in huts. Although the villagers are capable of repairing these houses by themselves, they are waiting for some organization to come forward to reconstruct new houses for them — an important observation that must be taken into consideration while considering villagers' ability to run a co-operative. CASA has given them wires to fence their drinking

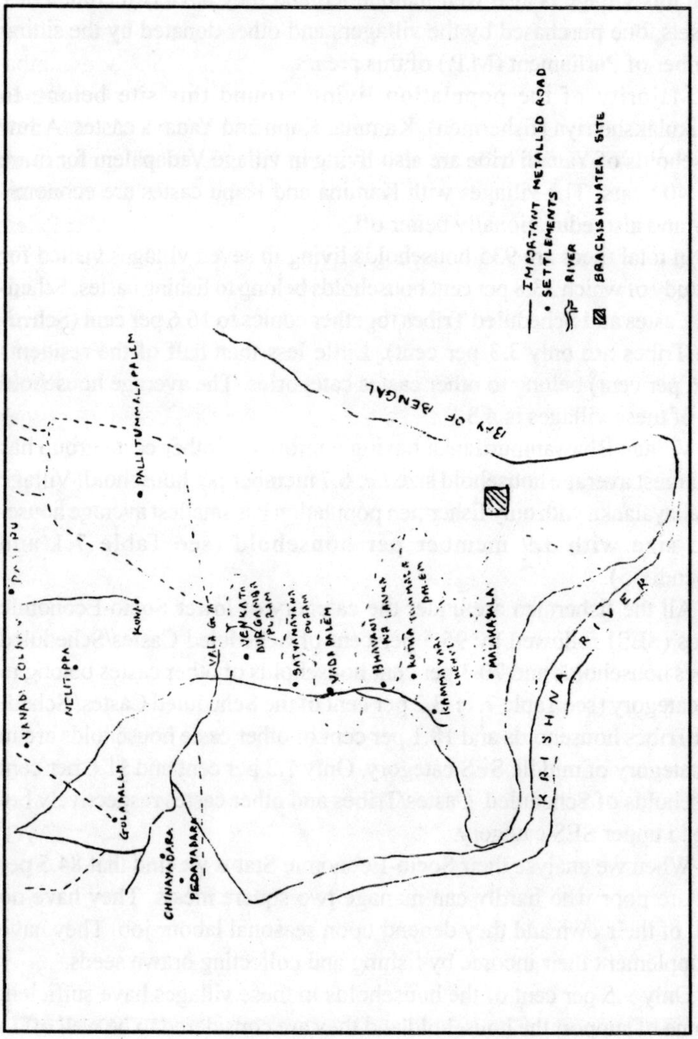

Map 7.3 : Map showing villages visited at P. T. Palem Brackish Water Site

water pond. It seems that they have developed a sense of dependence on outside help. The same type of one room kitchen house sets given to the villagers of Kammavaricheruvu are better maintained. The cyclone shelter in this village is also well maintained and they have two community TV sets, one purchased by the villagers and other donated by the sitting Member of Parliament (M.P.) of this area.

Majority of the population living around this site belong to Agnikulakshetriya (fishermen), Kamma, Kapu and Yadava castes. A few households of Yanadi tribe are also living in village Vadapalem for more than 40 years. The villages with Kamma and Kapu castes are economically and also educationally better off.

In total there are 935 households living in seven villages visited for the study of which 33.8 per cent households belong to fishing castes. Scheduled Castes and Scheduled Tribes together comes to 16.6 per cent (Scheduled Tribes are only 3.3 per cent). Little less than half of the residents (49.6 per cent) belong to other castes categories. The average household size of these villages is 4.3.

Village Bhawanipurilanka having a mjority of other caste group has the largest average household size *i.e.* 6.7 member per household. Village Malakayalanka with only fishermen population has smallest average household size with 3.9 member per household (see Table 7.1 and Appendix-5).

All the fishermen are under the category of lower Socio-Economic Status (SES) followed by 95.5 per cent of Scheduled Castes/Scheduled Tribes households and 70.3 per cent households of other castes belong to this category (see Table 7.1). 3.2 per cent of the Scheduled Castes/ Scheduled Tribes households and 18.1 per cent of other caste households are in the category of middle SES category. Only 1.3 per cent and 11.6 per cent households of Scheduled Castes/Tribes and other castes respectively belong to upper SES category.

When we analyse their Socio-Economic Status we find that 84.5 per cent are poor who hardly can manage two square meals. They have no land of their own and they depend upon seasonal labour job. They have to supplement their income by fishing and collecting prawn seeds.

Only 9.5 per cent of the households in these villages have sufficient income of support the household and they are considered to be well off in the village. Not more than 6 per cent of the villagers belong to the category of upper income group, those who have some surplus resources in terms of agricultural products or cash as saving.

As we move towards Chinnapuram or Machilipatnam, the

villages are developed in respect of their socio-economic condition. The villages in the P.T. Palem brackish water site area can be divided into two major types : (a) the village with Agnikulakshetriya (fishermen) caste and (b) villages having Kamma and Kapu as dominant castes.

Every village has one primary school and at least one cyclone shelter. All the villages are electrified. In village Vadapalem, there are two cyclone shelters, one constructed by the government and other by Satya Sai Organization. These cyclone shelters are double storeyed cubical or round *pucca* stuctures with electricity and doors and windows. Except in case of village Kammavaricheruvu and Vadapalem in all other villages the cyclone shelters are in bad shape — windows are broken and electricity wire hanging or open, power has been disconnected. These cyclone shelters are also used as community hall. It can accommodate at least 200 persons in any cyclone emergency.

For medical facilities, marketing and also for high school education the villagers have to go to Chinnapuram, a distance of 6 to 12 km.

Villages of K.P.T. Palem, Malakayalanka and Kammavaricheruvu do not have wells or hand pumps. The salinity of ground water is so high that wells and hand pumps fail to provide potable drinking water. The canal and creeks of river Krishna bring water to the village ponds and the villagers use this water for drinking and all other purposes.

The population of the villages surronding P.T. Palem site has three distinct occupational groups, the cultivators, fishermen and agricultural labourers. The cultivators are mostly from Kamma caste, the fishermen belonging to Agnikulakshatriya caste and the labourers come from the poorest cross section of these villages. As indicated in Table 7.1 that one-third of this population has small piece of land up to 2 acres but it does not provide enough to support the family hence, the major source of income for most of them is to work in the farm of fellow villagers or in construction or in Machilipatnam town. When buses ply, they go to Machilipatnam for job. They get Rs. 15 to 20 per day. Fishermen go to fish in Krishna and also in the Bay of Bengal. They do not have organized fishing but use small country boat for the river and large size country boat for the sea fishing. They get an average of Rs. 50 per day during the 5 months of the season. Quality fishes are taken by fish agents but the second grade fishes are sold by women by carrying it to the neighbouring villages. The rate for prawn and head off prawn of 3-4 inches length is Rs. 60 and Rs. 100 per kilogram respectively.

Table 7.1 : Distribution of Caste Category by Households, Population and Socio-Economic Status in Study Villages

(Site : P. T. Palem)

Caste Category	*Number of Household*	*Total Population*	*Average Household Size*	*Socio-Economic Status*		
				Lower	*Middle*	*Upper*
Fishermen	316 (33.8)	1350 (33.5)	4.3	316 (40.0) (100.0)		
Scheduled Castes/ Scheduled Tribes	155 (16.6)	654 (16.2)	4.2	148 (18.7) (95.5)	5 (5.6) (3.2)	2 (3.6) (1.3)
Other Castes	464 (49.6)	2026 (50.3)	4.4	326 (41.3) (70.3)	84 (94.4) (18.1)	54 (96.4) (11.6)
Total	935 (100.0)	4030 (100.0)	4.3	790 (84.5)	89 (9.5)	56 (6.0)

Table 7.2 : Percentage Distribution of Respondents' Occupation by Annual Income

(Site : P. T. Palem)

Occupation	*Annual Income*			*Total*
	< 3000	*3000-5000*	*Above 5000*	
Fishing	7 41.18 46.67	9 52.94 45.00	1 5.88 16.67	17 100.00 41.46
Cultivation	6 42.86 40.00	4 28.57 20.00	4 28.57 66.67	14 100.00 34.15
Other Occupation	2 20.00 13.33	7 70.00 35.00	1 10.00 16.67	10 100.00 24.39
Total	15 36.59 100.00	20 48.78 100.00	6 14.63 100.00	41 100.00 100.00

The fishes are sold mainly to the fish agents either coming to the village or the nearest road point. In case the road is not good for the fish lorry to come, the fisherman go to Chinnapuram to sell their fishes to the agents and from here it is taken to Machilipatnam by van. At Machilipatnam these are packed properly with required size packets and sent to various places like Calcutta, Madras, Kankinada, Visakhapatnam, Cochin and Hyderabad by road. The prawns are then transported to the foreign countries. The payment of fishes depends upon the relation of the fishermen or fishing group with the agent(s). It is daily, weekly or fortnightly payment. For regular suppliers, the payment is fortnightly. The major fish companies are Babji, Pandu, and Nageshwar Rao. At Machilipatnam there are three ice factories of 20 tones capacity and around 25 factories of 2 tones capacity. These factories cater to the need of fishing industires. During the fish crop season November and December and April and May — on an average 5 galvanised truck and 2 big trucks full of fish go to various places.

The entrepreneurs are interested in setting up ice factory near P.T. Palem area but because of problem in constant power supply, they concentrate in Machilipatnam town only.

Analysis of Individual Responses

Information collected from 41 individual households have been analysed. The analysis of their caste category by income in Table 7.2 shows that 85.4 per cent of the household are below the annual income of Rs. 5000 and among them are 36.6 per cent households whose income is less than Rs. 3000 per year. Only 14.6 per cent have annual income above Rs. 5000.

Occupationally these responses have been categorised in to three; in fishing (41.5 per cent), in cultivation (31.1 per cent), and in other occupation, mainly labourers (24.4 per cent). All the fishermen are in fishing and 25 per cent Scheduled Castes and Scheduled Tribes and 65 per cent of the other caste people are in cultivation (see table 7.3). Most of them (61.0 per cent) live in thatched huts and 31.7 per cent have *pucca* houses (see Table 7.4). These are the houses of Kammas from Bhawanipurilanka and Vadapalem and a few of the houses constructed by CASA in the village Kammavaricheruvu.

63.4 per cent of the respondents are illiterate and only 4.9 per cent are educated beyond primary level. Among the illiterates are all the Schedules Castes/Schedules tribes and 82.4 per cent of fishermen (see Table 7.5).

Table 7.6 indicates that most of the respondents who visualised the role of women in barckish water fisheries have preferred them to work for feeding the fish and guarding the fish pond.

As per the analysis given in Table 7.7 all the respondents from P.T. Palem, want to handle the pond independently. Although 75.6 per cent of the respondents are confident of taking care of their ponds by themselves, still 24.4 per cent required assistance from others. Other castes and Scheduled Castes/Scheduled Tribes respondents show higher level of confidence than fishermen (see Table 7.8).

Of those potential owners of the ponds, only 48.8 per cent of the respondents want to run the co-operatives with the help of the government agencies, mainly the Department of Fisheries, whereas a large number (51.22 per cent) do not want any government help (including the fisheries department) in co-operative affairs (see Table 7.9). However the analysis given in Table 7.10 clearly indicates that while all of the respondents are inclined to form co-operatives of pond owners, 23 per cent insisted to form co-operative of their own caste(s). More than half of the respondents from fishermen (52.9 per cent) want to have co-operative with their own caste members.

Table 7.3 : Percentage Distribution of Respondents' Caste Category by Occupation

(Site : P. T. Palem)

Caste Category	*Occupation*			
	Fishing	*Cultivation*	*Other Occupation*	*Total*
Fishermen	17			17
	100.00			100.00
	100.00			41.46
Scheduled Castes/ Scheduled Tribes		1	3	4
		25.00	75.00	100.00
		7.14	30.00	9.76
Other Castes		13	7	20
		65.00	35.00	100.00
		92.86	70.00	48.78
Total	17	14	10	41
	41.46	34.15	24.39	100.00
	100.00	100.00	100.00	100.00

Table 7.4 : Percentage Distribution of Respondents' Caste Category by House Type

(Site : P. T. Palem)

Caste Category	*House Type*			*Total*
	Thatched Hut	*Kuchcha*	*Pucca Having Thatched Roof/ Pucca*	
Fishermen	7	1	9	17
	41.18	5.88	52.94	100.00
	28.00	33.33	69.23	41.46
Scheduled Castes/ Scheduled Tribes	3		1	4
	75.00	—	25.00	100.00
	12.00	—	7.69	9.76
Other Castes	15	2	3	20
	75.00	10.00	15.00	100.00
	60.00	66.67	23.08	48.78
Total	25	3	13	41
	60.98	7.32	31.71	100.00
	100.00	100.00	100.00	100.00

Table 7.5 : Percentage Distribution of Respondents' Caste Category by Education

(Site : P. T. Palem)

Caste Category	*Education*				*Total*
	Illiterate	*Below Primary Level*	*Below Secondary Level*	*Beyond Secondary Level*	
Fishermen	14	3			17
	82.35	17.65			100.00
	53.85	33.33			41.46
Scheduled Castes/ Scheduled Tribes	4				4
	100.00				100.00
	15.38				9.76
Other Castes	8	6	2	4	20
	40.00	30.00	10.00	20.00	100.00
	30.77	66.67	100.00	100.00	48.78
Total	26	9	2	4	41
	63.41	21.95	4.88	9.76	100.00
	100.00	100.00	100.00	100.00	100.00

Table 7.6 : Percentage Distribution of Respondents' Caste Category by Women Help

(Site : P. T. Palem)

Caste Category	*Women Help*				
	Feeding	*Farm Guarding*	*Marketing*	*No Help-House Wife Only*	*No Response*
Fishermen	17 43.59	3 27.27	2 100.00		
Scheduled Castes/ Scheduled Tribes	4 10.26	2 18.18			
Other Castes	18 46.15	6 54.55			
Total	39 100.00	11 100.00	2 100.00		

'N' is more because of multiple choice.

Table 7.7 : Percentage Distribution of Respondents' Caste Category by Attitude Towards Handling of Pond

(Site : P. T. Palem)

Caste Category	*Pond Handling*			*Total*
	Independently	*On Fixed Wages*	*On Co-operative Basis*	
Fishermen	17 100.00 41.46			17 100.00 41.46
Scheduled Castes/ Scheduled Tribes	4 100.00 9.76			4 100.00 9.76
Other Castes	20 48.78 48.78			20 100.00 100.00
Total	41 100.00 100.00			41 100.00 100.00

Table 7.8 : Percentage Distribution of Respondents' Caste Category by Attitude Towards Pond Care

(Site : P. T. Palem)

Caste Category	*Pond Care*		*Total*
	Can Take Himself/ Herself	*Assistance Required*	
Fishermen	10	7	17
	58.82	41.18	100.00
	32.26	70.00	41.46
Scheduled Castes/ Scheduled Tribes	3	1	4
	75.00	25.00	100.00
	9.68	10.00	9.76
Other Castes	18	2	20
	90.00	10.00	100.00
	58.06	20.00	48.78
Total	31	10	41
	75.61	24.39	100.00
	100.00	100.00	100.00

Co-operatives and Co-operation

Kammaricheruvu has Milk Producers Co-operative Society organized under the banner of A.P. Dairy Development Federation. Federation milk collection van comes to this village on fair weather and collects milk or the agent collects and transports it by any possible means of transportation available at that time. The entries are made in a pass-book given to the member milk producers and payments are made every fortnight. Society is expecting to get a milk chilling plant. Because of this activity villagers are encouraged to keep milch cattle, mainly buffalo. CASA has initiated a women welfare society, namely "Sharada Mahila Mandal" established in the year 1983. There are more than 70 members from this village and from village Bhauvampuram. Nine women of this village are on the executive body of this Mandal. One sewing machine was provided by CASA to this Mandal. Another machine was provided by the Panchayat and the third was donated by a Madras based voluntary organization. Because CASA organized the training classes in the village, there are 10 personal machines pruchased by individual households.

Table 7.9 : Percentage Distribution of Respondents' Caste Category by Attitude Towards Running of Co-operative Society

(Site : P. T. Palem)

Caste Category	*Running of Co-operative*			*Total*
	With Govern-ment / Fish. Dept. Help	*Without Government Help*	*No Response*	
Fishermen	3	14		17
	17.65	82.35		100.00
	15.00	66.67		41.46
Scheduled Castes/ Scheduled Tribes	3	1		4
	75.00	25.00		100.00
	15.00	4.76		9.76
Other Castes	14	6		20
	70.00	30.00		100.00
	70.00	28.57		48.78
Total	20	21		41
	48.78	51.22		100.00
	100.00	100.00		100.00

Table 7.10 : Percentage Distribution of Respondents' Caste Category by Joining of Co-operatives with Own Caste/Other Caste Members

(Site : P. T. Palem)

Caste Category	*Joining of Co-operatives*			*Total*
	With Own Caste Members	*With Other Caste Members*	*No Response*	
Fishermen	9	8		17
	52.94	47.06		100.00
	50.00	34.78		41.46
Scheduled Castes/ Scheduled Tribes	1	3		4
	25.00	75.00		100.00
	5.56	13.04		9.76
Other Castes	8	12		20
	40.00	60.00		100.00
	44.44	52.17		48.78
Total	18	23		41
	43.90	56.10		100.00
	100.00	100.00		100.00

District Rural Development Agencey (DRDA), Machilipatnam has allotted 28 ponds of brackish water fisheries , one acre each, to 28 women of this village last year. Other 56 ponds are under construction which will be allotted to another lot of 58 people of this village after completion of construction. The selection of the beneficiaries for the first 28 ponds had been decided by the fisheries department without consulting the villagers or fishermen co-operative society. The society registered its protest and it has been agreed upon that the next 56 ponds will be distributed to those recommended by the society. On these ponds the concerned household has started prawn culture under the guidance of Assistant Director (Fisheries), DRDA, Machilipatnam. They calculate a net income of Rs. 5,000 per pond. It has helped boosting their economy. The pond owner has constructed a small hut near the pond and durng the harvest season the male members camp there. Rest of the times, they go for fishing. The women of these households help in mixing the feed, putting it in the tank, and also in taking care of the pond.

In Machilipatnam there is a registered body — The Bandar Taluka Brackish Water Fish Farmers Multipurpose Co-operative Society Ltd., registered in 1987. Society has 200 individual members with their share capitals of Rs. 100 each.

The co-operative society was formed to take advantage of the government land and subsidy. For the last 6 years the society is tyring to acquire government land on society's name for shrimp culture although the individuals have their own farms. In Krishna District, about 2,500 hectare of land are under prawn farming. In Bandar Mandal alone 450-500 hectare of land is used for brackish water fish farming. According to Mr. T. Ramamohan Rao, the Preseident of the society, there are almost 6000 hectare of land in Bandar which is suitable for brackish water fishery development.

It is reported that a close monitoring is required for the functioning of the co-operative groups. It happened that through society the village ponds are taken on a lease of Rs. 20,000 and given it to private entrepreneur on a premium of Rs. 25,000. The profit is shared by the co-operative members. Fisheries department keeps quite because of the pressure from interest group(s) and the politicians.

BRACKISH WATER SITE : C.G. PALEM

The brackish water site C.G. Palem is part of the village Chintayagaripalem (C.G.Palem), a hamlet of Devarampadu revenue village under Ongole

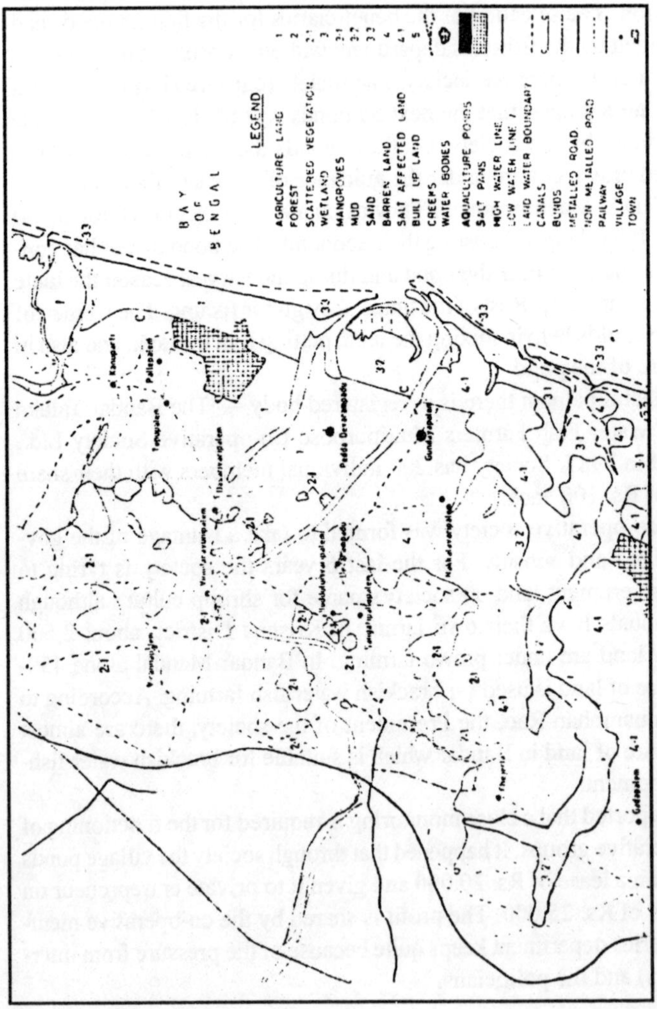

Source : Remote Sensing Application Group, ISRO, Ahmedabad, 1990.

Map 7.4 : Land Use Map Site : C. G. Palem

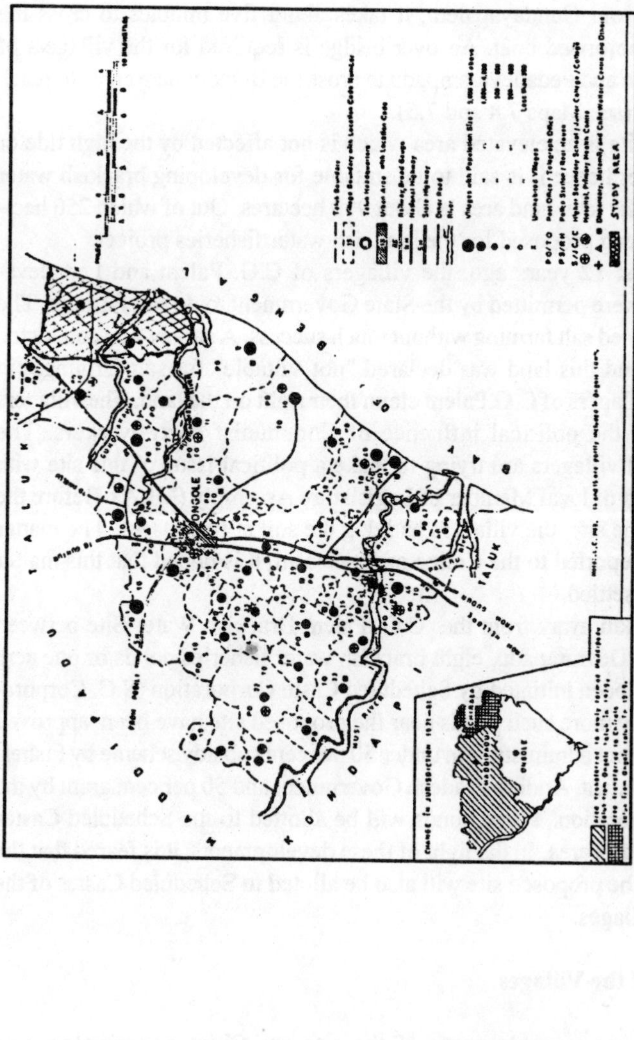

Based upon Survey of India Map with the permission of the Surveyor General of India. The Territorial States of India extended into the Sea to a distance of twelve nautical miles measured from the appropriate base line

Source : Census of India, 1981.

Map 7.5 : Ongole Taluk and Chirala Taluk, Prakasam District, Andhra Pradesh Site : C. G. Palem

taluk of Prakasam District. C.G. Palem is 25 km away from Ongole. The nearest railway station — Karvadi is 12 km. The site is located in an angular area surrounded by the Bay of Bengal from east by river Gundalakamma from south, and by Buckingham canal towards its west. North to the site is barren land. It is about half a kilometre from the village Gundayapalem and one km from the main village, C.G. Palem. The villagers living in this area have to either cross the canal or the river to reach the site. From Gundayapalem, it takes about five minutes to cross the river by motorised boat. An over bridge is required for the villagers of C.G. Palem and Pedadevarampadu to cross the Buckingham canal to reach to the site (see Maps 7.4 and 7.5).

The site is an elevated area which is not affected by the high tide or flood water hence it is said to be suitable for developing brackish water fisheris. The total land area is about 900 hectares. Out of which 250 hectares has been proposed for the brackish water fisheries project.

Almost 12 years ago, the villagers of C.G. Palem and Pedadevarampadu were permitted by the State Government to do salt farming. The villagers tried salt farming without much success. A survey was conducted in 1986, and this land was declared "not suitable" for salt farming, but still the villagers of C.G.Palem claim their right on the land. The villagers are under the political influence of Communist Party workers. The influencial villagers are trying to make a political issue of this site with the support of local Member of Legislature Assembly (MLA). Before the research the site, the villagers had dug the soil at two places. The matter has been reported to the Collector, Prakasam, it is hoped that this matter would be settled.

Four km away from the C.G. Palem Brackish Water Site between Ulchi and Devrampadu, eight brackish water fisheries ponds of one acre each have been initiated by Scheduled Caste Corporation (S.C. Corporation). Forty more such ponds near the proposed site have been approved by the district administration under 50 per cent subsidy scheme by Fisheries Department, Andhra Pradesh Government and 50 per cent grant by the S.C. Corporation. These ponds will be allotted to the Scheduled Castes people of this area. In the light of these developments, it is feared that the ponds of the proposed site will also be alloted to Scheduled Castes of the nearby villages.

Profile of the Villages

The villages towards north of the site are Chintayagaripalem and

Pedadevarampadu. Towards south are Gundayapalem and Pathapadu villages and towards west of the site on Gundayapalem — Ongole road are Sankatapallipalem, Harijanpadu and Devarampadu settlements. All these are the hamlets of the revenue village Devarampadu. Gundayapalem and Chintayagaripalem are mainly fisherman's villages. Sankatapallipalem and Harijanpadu are the villages of Scheduled Castes and Rajaka castes, whereas, Pedadevarampadu has only *Yadava* caste living there. Pathapadu and Devarampadu villages are dominated by other castes such as *Kamma* and *Kapu.*

The approach to the C.G. Palem brackish water site is through fair weather metaled road reaching up to the village Gundayapalem. Andhra Pradesh State bus comes up to Gundayapalem at least four times a day during the fair weather (about six months in a year). On this route, there are four places which can be characterized as "low lying" where most of the time there is waterlogging up to 1.5 feet. Villagers of C.G. Palem go 3 km north to take a bus from Desarivaripalem, Chirala *taluk.*

Pathapadu is difficult to reach and the waterway is the only suggested route. The motorised boat may reach up to the village only during the high tide depending upon the level of water in the Buckingham canal. The village has a wooden bridge to cross the canal and a seasonal *kuchcha* road to reach to the Karvadi village which is only seven km from Patapadu. This road connects Patapadu to the railway station and Ongole. *Kamma* caste dominates the village and is economically better off in the sense that in a village of 118 households, there are three individually owned TV sets and approximately 70 transistor sets apart from seven spacious houses having brick walls and tiled roof. Fourty-eight standard size *pucca* houses have been built under Housing Corporation subsidy scheme. The soil of Patapadu is also suitable for groundnut cultivation. In the last five years, villagers have grown groundnut on 30 acres of land.

As shown in the Map 7.6 the village settlements around the brackish water site C.G. Palem are situated either by the side of the river Gundalakamma or by the side of Buckingham canal. Most of these settlements , referred to as "villages" are the hamlets of the revenue villages. All the villages locally referred to as Panchayat for C.G. Palem site are the part of Devarampadu revenue village.

In seven selected villages for the study live 1436 households and the estimated population is 5623 which give an average household size of 4.6 members. Village Devarampadu has the largest average size having 8 members per household. Fishermen form the majority of villages having 61.9 households followed by 29.1 per cent other castes and 16.9 per cent Scheduled Castes. Apart from the fishermen castes, the other castes also

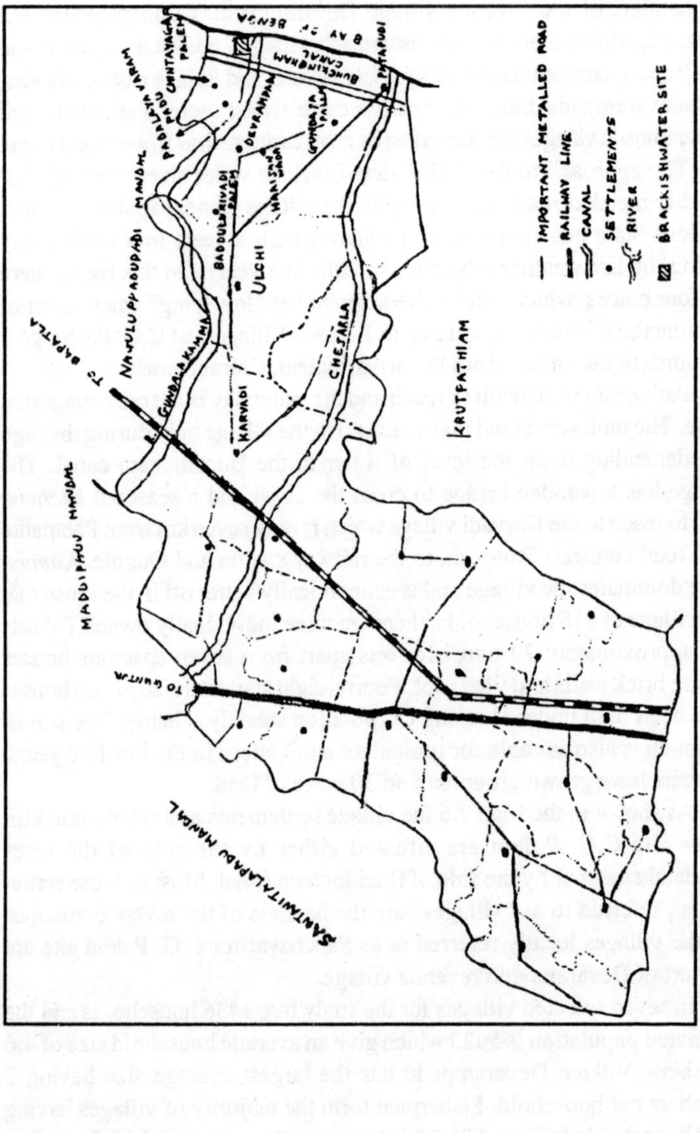

Map 7.6 : Map showing villages visited at C. G. Palem Brackish Water Site

do fishing (nearly 40 per cent of them) to support their economy. All the Scheduled Caste population and quite a sizable number of other castes are involved in labour work. A few castes like Rajaka, Kamara and Swarankara are in their traditional occupation of washing clothes, blacksmithy, and goldsmithy respectively.

The socio-economic condition of these villagers is the reflection of their plight. 98.1 per cent of the villagers are poor and only 1.9 per cent of them have sufficient to support their families. None of the villagers were said to be in the upper income group who have more than what they require for maintaining their households. Although the villagers of Devarampadu look little affluent with their *pucca* houses and better looking living condition, but since they only depend upon agriculture, they claim that after investing as input to the agriculture and after repaying the loans they hardly get enough to survive. Because of the prestige attached to their castes, they do not even work as labourer.

The condition of the Scheduled Castes and *Rajakas* (the washermen caste) are pathetic. Fishing is the only source of income for the fishermen of these villages. With the help of their co-operatives, they have been able to acquire nets and boats for fishing in the river and the sea but the rest of the poor population have to work as daily wage labourers. Most of them work as *koolies* (labourers) in the salt farm where they get up to Rs. 15 for men and Rs. 12 for women per day. This employment is available only for 45 to 60 days in a year. Rest of the time they go to Ongole to work.

During agriculture season they get employment in transplantation as well as in harvest of paddy. The *Rajaka* women are in their traditional caste occupation of washing clothes for the villagers. They get both cash and crop but this is just not enough to support the family. The statement by one *Rajaka* woman from Sankatapalli — "we get left overs from three-four client houses and feed our children" — is a representative statement of the 20 *Rajaka* households in the village. The Yadavas of Pedadevarampadu are slightly better and have more land but they are in search for an alternative source of income.

All the villages have a primary or lower primary school. The nearest high school is at Karavadi. For medical treatment also, villagers go to Karavadi. Because of better bus communication, most of them prefer to go to Ongole of treatment.

Gundayapalem is a fairly large village. It is the centre for fishing activities because it is the last bus stop and is situated at the bank of the river. Fish collection vans come up to this village during fair weather, otherwise the villagers carry the fish load up to Ulchi by bullockcart. Trucks also come to the village for transporting fish as well as salt produced in this area.

Devarampadu, Harijanpadu, and Gundayapalem have a few small shops to meet the daily minor needs. For major purchase and other requirements, the villagers go to Ulchi or Ongole.

With exception of the houses constructed under the subsidy scheme, all houses are *kuchcha* houses (22.5 per cent) or thatched (60.0 per cent) houses. Gundayapalem and Patapadu villages have standard size (one room one kitchen) houses built by life Insurance Corporation under Housing Corporation Scheme of 50 per cent subsidy in 1983. In Gundayapalem alone there are 400 such houses. By now most of the houses are leaking. In another scheme Rs. 7000 per house was provided by the government to build houses in 1990. Ten such houses are under construction and 40 more will be taken up in the next financial year.

The fish companies operating from Ongole come to the village for fish collection. They offer interest free advance to potential fishermen. Almost 50 households from among the 700 fishermen caste households have borrowed money from the fish companies of fishermen in order to meet their household requirements. The borrowed money has been used for buying nets, boats repair and for purchase of bullockcart. The borrowed money is paid back through fish sale of a discount rate of 20 per cent from the usual market price. These fishermen are morally bound to sell their fish to the advancing agency. It also happened in the past that advance was taken from one agent but fish was sold to another on cash payment. These litigations are settled by the ward representatives and a few fellow villagers interfering in it.

Each fisherman household earn on an average Rs. 30 to 50 in the fish season after paying the installment. The season in 5-6 months a year. For 2 months they get prawn and the income is up to Rs. 70 to 80 per day. The women of these household sell fishes in nearby villages. The fish companies collect only quality fishes. These women also some time help in collecting natural prawn seeds and sell it to the interested parties.

Analysis of Individual Responses

Analysis of 40 responses collected from 40 individual households indicate that majority of the respondents live below poverty line. Only five per cent have indicated their annual income above Rs. 5000, whereas, 22.5 per cent have an annual income less than Rs. 3000. A large part of the respondents (72.5 per cent) get an annual income between Rs. 3000 and Rs. 5000 (see Table 7.1).

When we analyse the caste categories by their occupations in Table 7.13 , we find more or less equal distribution in occupation, *i.e.*, fishermen (30.0 per cent), cultivators (37.5 per cent) and other occupations

Table 7.11 : Distribution of Caste Category by Households, Population and Socio-Economic Status in Study Villages

(Site : C. G. Palem)

Caste Category	*Number of Household*	*Total Population*	*Population Household Ratio*	*Socio-Economic Status*		
				Lower	*Middle*	*Upper*
Fishermen	893 (61.9)	4003 (59.9)	4.5	893 (63.1) (100.0)		
Scheduled Castes/ Scheduled Tribes	231 (16.0)	841 (12.6)	3.6	231 (16.3) (100.0)		
Other Castes	319 (22.1)	1837 (27.5)	5.8	292 (20.6) (91.5)	27 (100.0) (8.5)	
Total	1443 (100.0)	6681 (100.0)	4.6	1416 (98.1)	27 (1.9)	

Table 7.12 : Percentage Distribution of Respondents Occupation by Annual Income

(Site : C. G. Palem)

Occupation	*Annual Income*			*Total*
	< 3000	*3000-5000*	*Above 5000*	
Fishing	3 25.00 37.50	8 66.67 27.59	1 8.33 50.00	12 100.00 30.00
Cultivation	2 13.33 25.00	12 80.00 41.38	1 6.67 50.00	15 100.00 37.50
Other Occupation	3 23.08 37.50	9 69.23 31.03		13 100.00 32.50
Total	8 20.00 100.00	29 72.50 100.00	2 5.00 100.00	40 100.00 100.00

Table 7.13 : Percentage Distribution of Respondents' Caste Category by Occupation

(Site : C. G. Palem)

Caste Category	*Occupation*			*Total*
	Fishing	*Cultivation*	*Other Occupation*	
Fishermen	12 75.00 100.00	3 18.75 20.00	1 6.25 7.69	16 100.00 40.00
Scheduled Castes/ Scheduled Tribes		3 42.86 20.00	4 57.14 30.77	7 100.00 17.50
Other Castes		9 52.94 60.00	8 47.06 61.54	17 100.00 42.50
Total	12 30.00 100.00	15 37.50 100.00	13 32.50 100.00	40 100.00 100.00

Table 7.14 : Percentage Distribution of Respondents' Caste Category by House-Type

(Site : C. G. Palem)

Caste Category	*House-Type*			*Total*
	Thatched Hut	*Kuchcha*	*Pucca Having Thatched Roof/ Pucca*	
Fishermen	13 81.25 54.17	2 12.50 22.22	1 6.25 14.29	16 100.00 40.00
Scheduled Castes/ Scheduled Tribes	4 57.14 16.67	2 28.57 22.22	1 14.29 14.29	7 100.00 17.50
Other Castes	7 41.18 29.17	5 29.41 55.56	5 29.41 71.43	17 100.00 42.50
Total	24 60.00 100.00	9 22.50 100.00	7 17.50 100.00	40 100.00 100.00

dominated by labourers which comes to 32.5 per cent. Obviously, all the fishermen have indicated fishing as their main occupation. The labourers in this area work in salt farms. Majority of them (60 per cent) live in thatched huts and 22.5 per cent live in *kuchcha* houses of mud walls and tiled roof. 12.5 per cent have *pucca* houses (see table 7.14). These respondents are from Devarampadu and Pathapadu villages belonging to Kamma castes. Two respondents living in the housing board houses also come in this category. The analysis provided in Table 7.15 shows more than three-fourth (77.5 per cent) are illiterates and only one of the respondents is educated beyond secondary level. Only ten per cent of them have studies above primary level of education.

The analysis of multiple choice answers about the role of women would play in brackish water fisheries given in Table 7.16 shows that women's role is associated with farm activities like guarding the ponds as well as preparation of fish feed and in feeding them to fishes. Marketing of fish was indicated by the fishermen who are already practicing it and lone household from Scheduled Castes.

All the 40 respondents expressed their opinion that they would like to handle the pond independently and do not want to work on fixed wage. (see Table 7.17). Although 70 per cent of the respondents are confident of taking care of their ponds themselves 30 per cent required assistance from others. The Scheduled Castes/Scheduled Tribes and fishermen respondents have higher degree of confidence (100.0 per cent and 68.8 per cent respectively), only 58.8 per cent of the respondents from other castes indicated that they can take care of an acre pond themselves (see Table 7.18). 60.0 per cent of the respondents want to run the co-operatives with the help of the government agencies, mainly the Department of Fisheries, and 40.0 per cent did not want any government help including from the fisheries department (see Table 7.19).

On the question of forming a co-operative with other pond owners the respondents expressed their different views. The analysis given in Table 7.20 indicates that while all of the respondents are inclined to from co-operatives of pond owners. 7.5 per cent insisted to form co-operative of their own caste(s). All the respondents belonging to Scheduled Castes/Scheduled Tribes expressed no reservation. The personal interviews with them revealed that it is not the Scheduled Castes/Scheduled Tribes but the other caste members would not like to join co-operatives with so called lower castes.

Co-operatives and Co-operation

Gundayapalem, Chintayagaripalem and Pathapadu villages are the main

Table 7.15 : Percentage Distribution of Respondents' Caste Category by Education

(Site : C. G. Palem)

Caste Category	*Education*				*Total*
	Illiterate	*Below Primary Level*	*Below Secondary Level*	*Beyond Secondary Level*	
Fishermen	10	3	2	1	16
	62.50	18.75	12.50	6.25	100.00
	32.26	60.00	100.00	50.00	40.00
Scheduled Castes/ Scheduled Tribes	7				7
	100.00				100.00
	22.58				17.50
Other Castes	14	2	—	1	17
	82.35	11.76		5.88	100.00
	45.16	40.00		50.00	42.50
Total	31	5	2	2	40
	77.50	12.50	5.00	5.00	100.00
	100.00	100.00	100.00	100.00	100.00

Table 7.16 : Percentage Distribution of Respondents' Caste Category by Women Help

(Site : C. G. Palem)

Caste Category	*Women Help*				
	Feeding	*Farm Guarding*	*Marketing*	*No Help-House Wife Only*	*No Response*
Fishermen	16	16	4		
	64.00	64.00	80.00		
Scheduled Castes/ Scheduled Tribes	6	7	1		
	24.00	28.00	20.00		
Other Castes	3	2			
	12.00	8.00			
Total	25	25	5		
	100.00	100.00	100.00		

'N' is more because of multiple choice.

Table 7.17 : Percentage Distribution of Respondents' Caste Category by Attitude Towards Handling of Pond

(Site : C. G. Palem)

Caste Category	*Pond Handling* *Independently*	*On Fixed Wages*	*On Co-operative Basis*	*Total*
Fishermen	16			16
	100.00			100.00
	40.00			40.00
Scheduled Castes/ Scheduled Tribes	7			7
	100.00			100.00
	17.50			17.50
Other Castes	17			17
	100.00			100.00
	42.50			42.50
Total	40			40
	100.00			100.00
	100.00			100.00

Table 7.18 : Percentage Distribution of Respondents' Caste Category by Attitude Towards Pond Care

(Site : C. G. Palem)

Caste Category	*Pond Care* *Can Take Himself/Herself*	*Assistance Required*	*Total*
Fishermen	11	5	16
	68.75	31.25	100.00
	39.29	41.67	40.00
Scheduled Castes/ Scheduled Tribes	7		7
	100.00		100.00
	25.00		17.50
Other Castes	10	7	17
	58.82	41.18	100.00
	35.71	58.33	42.50
Total	28	12	40
	70.00	30.00	100.00
	100.00	100.00	100.00

Table 7.19 : Percentage Distribution of Respondents' Caste Category by Attitude Towards Running of Co-operative Society

(Site : C. G. Palem)

Caste Category	*Running of Co-operative*			*Total*
	With Government / Fish. Dept. Help	*Without Government Help*	*No Response*	
Fishermen	11 68.75 45.83	5 31.25 31.25		16 100.00 40.00
Scheduled Castes/ Scheduled Tribes	5 71.43 20.83	2 28.57 12.50		7 100.00 17.50
Other Castes	8 47.06 33.33	9 52.94 56.25		17 100.00 42.50
Total	24 60.00 100.00	16 40.00 100.00		40 100.00 100.00

Table 7.20 : Percentage Distribution of Respondents' Caste Category by Joining of Co-operatives with Own Caste/Other Caste Members

(Site : C. G. Palem)

Caste Category	*Joining of Co-operative*			*Total*
	With Own Caste Members	*With Other Caste Members*	*No Response*	
Fishermen	1 6.25 33.33	15 93.75 40.54		16 100.00 40.00
Scheduled Castes/ Scheduled Tribes		7 100.00 18.92		7 100.00 17.50
Other Castes	2 11.76 66.67	15 88.24 40.54		17 100.00 42.50
Total	3 7.50 100.00	37 92.50 100.00		40 100.00 100.00

fishermen villages and all of them have fisherman Co-operative Society formed almost 10 years ago. In all these villages, there are members of Salt Manufacturers' Co-operative Society. In addition to this, in Gundayapalem village there is a co-operative soceity for casurina plantation on a small piece of land given to them by the government.

Since 1975, Gundayapalem has an active fishermen co-operative society with 250 members. This type of societies also exist in Pathapadu and Chintayagaripalem. These societies are registred with fisheries department who not only facilitate the societies in terms of processing their papers for land lease, loans, and getting boats and nets for them on subsidised rates, it also conduct election of President and Executive Committee members every three years. In Gundayapalem there are 50 boats, 25 bought by individuals and 25 by co-operative efforts on loan. The co-operative society prepares a list of applicants and make a group of six individuals from six households selected by the society, who apply for the loan to buy boats. This group of six use the boat for fishing and the catch is sold by the individual made responsible for it. A fixed amount of Rs. 50 every catch is kept for the repayment to the loaning agency and rest is shared equally by them. This group is responsible for the maintenance of the boat. In case of conflicts the villagers call a Panchayat meeting and an agreed amount is paid to the individual and he is expelled out. The amount is calculated on the basis of the present cost of boat, the repayment of installments and the expected benefit the individual has already got out of the boat. The cost of the ordinary country boat is Rs. 25,000 and there is a 33 per cent subsidy on it. Beach Landing Craft (BLP), the motorized fiber boat is costing Rs. 1,30,000 and there is no subsidy on it. The loans are obtained mostly by state co-operative department on nine per cent interest rate or from commercial bank on 12.5 per cent interest rate. The repayment tenure is five years. There are five BLPs in Gundayapalem. Fishermen are not very enthusiastic about BLP scheme for cost and maintenance problem. There is no repair facility in Ongole and they have to call the mechanic from Nellore which is expensive as well as time taking and till then the boat will be anchored on the shore.

REFERENCES CITED

Draft Preparation Report. Brackish Water and Inland Fisheries Project, India 1989. Rome : Investment Centre, FAO/World Bank Co-operative Programme.

Molnar, Augusta, 1989 Community Forestry : Rapid Appraisal, A Review Paper, Rome : Food and Agriculture Organization of the United Nations.

APPENDICES

APPENDIX-1

7 November, 1990

DRAFT PROPOSAL FOR SOCIO-ECONOMIC STUDY

(Dr. Binod C. Agrawal)

INDIA

BRACKISH WATER AND INLAND FISHERIES PROJECT

Brackish Water and Inland Fisheries Project : Socio-Economic Aspets for Study

Based on the analysis presented and questions raised in Annex-3 of the Draft Preparation Report, the following sociological issues emerged for further study. The answer to these questions will help decide on organizational and management structures that should be proposed for the implementation of the project on semi-intensive shrimp production in the selected sites in India.

1. What are the demographic and socio-economic characteristics of the potential beneficiaries ?
2. What kind of consultations with the potential beneficiaries (the poor and the weaker sections of the society living in the coastal villages around the proposed project sites) should be arranged to decide the most preferred arrangements for farm operation and management ?
3. What should be the criteria for the selection of beneficiary villages near the project sites ?
4. What are the existing farm organizations for brackish water fisheries and other fisheries ?
5. Are there existing "Co-operative models" being successfully practiced

in other areas like "credit, "dairy", "sugarcane", *"bidi"*, "weaver" and similar production systems. Analysis of this kind will help design the "Co-operative model" for the brackish water fisheries project.

6. What are the village organizations that are prevalent in three States of Andhra Pradesh, Orissa and West Bengal near proposed sites ?
7. What are the organizational options considered most suitable by the officials of each state ?
8. How one selects a beneficiary so as the fruits of the Brackish Water Fisheries reach to the poor of the villages located near five kilometres of the proposed sites ?
9. How one designs the entire system by which all aspects of the project move in the desired direction ?
10. What are the different approaches that must be followed to break away from the past philosophy of "subsidy" and "financial support" provided by the Government of India/State Governments to initiate and run such projects/schemes like IRDP and BFDA ?
11. What are the technical and organizational training required at the farm level (both for the individuals and groups/organizations) to run the farm operations smoothly.

Answers of these Questions will help Decide at Farm Level :

(i) How to involve the potentially poor individual beneficiaries in the farm operations and mangement to provide maximum benefits to them?

(ii) What size of ponds should be designed for a given farm?

(iii) What kind of techno-managerial support should be provided whether it be state, consultancy groups, sprearhead teams, techno-managerial teams appointed/hired by the individual co-operatives?

(iv) Who would be responsible for masterminding the total operation, input supply, production, supervision and marketing of individual sites in a given state?

(v) How the benefits/losses will be shared by the individual, community, and co-operatives, or even state?

13. In order to answer these questions, it is suggested that with the help of Rapid Rural Appraisal (RRA) technique being followed in several parts of the world, first hand data from the selected proposed sites should be collected and analysed. In addition, maps should be developed showing

the position of the villages in relation to the project sites using thematic maps prepared by ISRO falling within a radius of five kilometres. These maps should incorporate information about the current status of the village (inhabited or uninhabited), transport, marketing, communication networks. All these data should be verified from the field.

14. Given the time constraints, it may not be possible to visit all 17 proposed site for RRA. Hence, it is proposed that two sites, in order of preference, based on proposed sechedule of development in Annex 1, Figure 8, will be visited where firsthand data will be collected on the issues raised earlier. As physical size of the site determines the number of beneficiaries, it is proposed that the investigation will follow a purposive sampling technique. For this purposive sampling, five "concentric circles" of one kilometre around each proposed site, and first all the villages falling in the first concentric will be covered. This process of investigation will continue until three times more of the required number of households on the proposed site is covered. Also efforts will be made to locate geographically suitable villages, which do not fall within the five kilometre circle though logistically suitable for the project. In the case of remaining sites, data on the proposed proforma No. 1 (Annex 1) will be collected with the help of Remote Sensing Ground Truth Team of ISRO or state and district officials as far as possible.
15. In order to provide specific answers to some of the important issues for planning decisions, few local areas will be investigated. These are discussed in the next section.

Focal Areas of Investigation

(i) Co-operatives of "sugarcane" and "dairy" and existing research studies on Co-operatives at the Institute of Co-operative Management at Pune will be studied. This will provide qualitative insights to understand and to reach some definitive conclusions on the kind of organization and mangement structure that should be followed, bearing in mind the socio-political implications of the suggested models.

(ii) Preparation of thematic maps of all the proposed sites within a five kilometre radius. These will be verified during the field visits with the help of district and block officials.

(iii) Actual verification of six thematic maps for locating the habited villages by the field team during RRA study and visits.

(iv) Demographic, socio-economic, landholding, occupational and religious data will be compiled from the existing census handbooks, District Mandal and Block Level Offices. In the case of the six selected sites for RRA all efforts will be made to update this information during the village visits.

(v) Infrastructural facilities available within the five kilometre radius of each site, including the marketing centres and other centres having larger/bigger infrastructure facilities will be collected.

(vi) Specific data on "fishing communities" living within the five kilometre radius — current fishing practices, ownership of fishing boats, leadership pattern, their views and opinions about the "ideal" or "workable" methods of organizing the training, management and organization of the fish farms within their area, focused investigation on the level of "community cohesion" belief in co-operation, trust/distrust for outside persons including government officials.

(vii) Attitude of the villagers (both men and women) and fishing community towards acceptance of brackish water fisheries, farm operations, co-operation, community participation, leadership, Government involvement in the mangement of the fish farms, aceptance of non-subsidy scheme, and training and re-training of the fishermen/women to participate in the proposed project. Also efforts will be made to investigate the perceived economic advantages/dis-advantages and conflicts arising out of this project, implementation at various levels within the village and in surrounding areas. Efforts will be made to collect data both from men and women separately.

Site Selection of Rapid Rural Appraisal

17. Based on several discussions with the FAO Project Mission Team, state officials and available literature on various proposed sites, it is thought most appropriate to select at least one site from each of the three States, namely Andhra Pradesh, Orissa and West Bengal. All three States are divergent in terms of the language they speak, (Telegu in Andhra Pradesh, Oriya in Orissa and Bengali in West Bengal); the experiences gained in land reforms, and to some extent current political philosophy followed in each State. While coastal population in each state might be more or less at the same level of economic development, there are visible differences among

the States. Hence, one site in each State should be selected for RRA. It would be desirable to have at least two sites in each State for RRA study, hence, it is proposed to investigate two sites. The selection is based on the order of preference of the project implementation indicated in Figure 8.

1. P.T. Palem (Andhra Pradesh)
2. C.G. Palem (Andhra Pradesh)
3. Jagatjore (Orissa)
4. Panaspada (Orissa)
5. Digha (West Bengal)
6. Dadanpatra (West Bengal)

It may be kept in view that if due to some unforseen reasons it would not be possible to visit these alternate sites from among the proposed sites in each state will be studied following the order of preference indicated in Figure 8.

(i) For each site, with the help of thematic maps and village maps available in the *District Census Handbook of 1981* all the habited villages falling within five kilometre radius will be identified and listed. All available secondary data available in the government records at the district level for each village will be collected.

(ii) On the same map, for each site, all major roads, marketing centres, communication links, and other infrastructural facilities will be plotted. As far as possible, a short description of the linkages, availability of services and goods in these areas will also be provided. While doing so, one of the major efforts will be to indicate current state of fish related activities in the area.

(iii) General description on the village social structure, occupation, leadership, economic condition of the people and social stratification of the area will also be provided. Special focus will be drawn on the fish related co-operatives, if any, working in the area and the experiences of the people in this regard.

(iv) In the case of the six selected sites for RRA study, after completion of the first three items indicated earlier, field visits will be made. Efforts are being made to select three "project leaders" of ISRO who alongwith about 4 field investigators will spend about four weeks time in each state, while the project leaders will be part of the DECU/ISRO senior social science group, the field investigators will be locally hired having Masters/Bachelors degree in social science having reading, writing and speaking ability of the local language. This will cut down on the data collection time and also help establish rapport with the villagers. The filed team with the help of the logistic support to be provided by the State Fisheries Department will collect field data following the method suggested in Section 14 of this proposal.

(v) Three kinds of data collection will be collected in the field. These are :

(a) Village level data;
(b) Group level data including fishermen caste level data; and
(c) Individual level data.

(vi) **Village Level Data** : Detailed demographic, socio-economic and occupational data will be collected from the key informant, village leaders, village officials and block officials (see Annex 1). By triangulation or cross-checking, an effort will be made to get the correct information. A "Village Schedule Guide" (see Annex 2) has been prepared for this purpose. The final version will be further enlarged and will be uniformally used. Efforts will be made to have special observations, comments and remarks by the field investigators and project leaders. For each village this data will be separately collected. During the visit of the village efforts will be made to write about first hand observation of the research team members about the village as a whole and all the developmental activities running in the village and their level of success. Also case studies (at least one from each) will be prepared based on the outline given in Annex 3.

(vii) **Group Level Data :** With the help of an open ended: "Interview Schedule Guide" (see Annex 4) the teams will try to gather fishing community in the village, and hold detailed discussions lasting, may be, more than an hour. During this group discussion, separate groups of men and women will be drawn into discussing about various issues relating to brackish water fisheries after giving them details about the proposed brackish water fisheries project proposal before initiating the discussion. Depending upon the situation, all efforts will be made so that the local officials of the Fisheries Department can be present during these discussion sessions to give credence to the fact that this project is under serious consideration by the Government and the same turn into reality for the prosperity and wellbeing of the people of the village. Issues raised regarding type of preferred organization of the project, attitude of the fishing and poor communities will also be ascertained. During these discussions questions will be raised about the existing development schemes and the chances of the acceptance of the proposed one. The research team with the help of local leaders and officials will identify poor and weaker sections while collecting data on Proforma 1, Annex 1. Also during these group discussions three fishing families, and poor non-fishing households will be indentified for individual interviews. Among three at least, one will be selected having brackish water fisheries experience.

(viii) **Individual Household Data :** Two kinds of household from each village will be selected for collecting household data (Annex 4 not appended but will be developed as discussed). These will be:

(a) Fishermen household including one with brackish water fisheries experience; and
(b) Poor but non-fishermen household.

In each village at least three households from each category will be selcted as discussed earlier, and interviewed. Analysis will be done at an aggregate level for a single site and also for all the three sites. Some attitude questions will be asked to the respondents alongwith their socio-economic and educational background data. Keeping in view that the decision-making power lies with the head of the household in each of three selected sites. The head of the household will be the respondent for this part of the study. A semi-structured interview schedule guide will be developed in English with translations in each of the three languages (see Annex 5). Data collection will be done in the local language, though the responses will be noted as far as possible in English. Some household of level observations will be also made about the respondent's perception about the chances of the acceptance and success of non-subsidy based development scheme like the one suggested now.

Also, the extent to which the respondent will be willing to participate in this project having no subsidy.

REPORT

18. Village level data analysis will be done in the village every day. However, data integration will be done centrally at Ahmedabad and the report will be prepared having an introductory Chapter 1; Methodology Chapter 2; three levels of analyses in Chapters 3, 4, 5; Integration of three levels of analyses in Chapter 6; and final chapter with recommendations and implications of the study for this project in Chapter 7. First effort will be to bring out the main findings and recommendations of the study to the mission team.

Proposed Timetable for the Study

1. Initial preparation of :
   (a) thematic maps at Ahmedabad for all the proposed sites; and
   (b) preparation of data collection instruments and preparation of the field visits.
   Two weeks *26 November - 9 December, 1990*
   (Thematic mapping will continue if needed).

2. Visit to the field (one project leader per state) alongwith one Remote Sensing Expert for ground truth (this will be decided only after return to Ahmedabad and talking to Dr. Baldev Sahai. In the field:
   (a) recruitment of four (two male and two female) field investigators from the local areas;
   (b) training of the field investigators;
   (c) site visits and data collection from all the villages having habitation;
   (d) visit to officials of the Fisheries Department for checking that data of other sites and for discussion and cross-checking of data regarding other development schemes running in the area.

— *Four weeks (10 December, 1990 — 7 January, 1991)*

3. Data analysis and report writing at Ahmedabad.

— *Two weeks (8 January — 21 January, 1991)*

4. Submission of the preliminary report to FAO, Rome and Ministry of Agriculture, New Delhi.

— *One week (21-27 January, 1991)*

A more detailed work plan will be evolved and sent to Government later.

The above plans are based on the assumption that there will be no objection raised by the Government of India, respective State Governments and ISRO authorities. Given the time constriants co-operation of the State officials will be essential in achieving the task.

ANNEX 1

Proforma

1. Village Level Data on Caste, Occupation and Economy
2. Name of Site
3. Name of the Village
4. Name of Taluka/Mandal/Block
5. Name of the District
6. Distance of the Site from the Village in kilometres

S. No.	Name of the Caste (Religion)	Total No. Household	Occupation	Whether SC/ST	Economic Status	Whether involved in brackish water fisheries	How many
1							
2							
3							
4							
5							
6							
7							
8							
9							
10							
11							
12							
13							
14							
15							
16							
17							
18							
19							
20							
Total							

ANNEX 2

Village Schedule Guide

1. Name of the site : State
2. Name of the Village.
3. Name of the Taluka/Mandal/Block.
4. Name of the District.
5. Is this is permanent/semi-permanent/seasonal settlement village?
6. What is the estimated population of the village?
7. What is the religious/caste composition of the village? (Check from Proforma 1).
8. What are the major occupations in the village ? (Special efforts should be made to check on first related activities).
9. What kind of leadership exists in the village :

   (a) Traditional (caste based);
   (b) Traditional (village based);
   (c) Panchayat or elected leadership?

10. Who are the economically powerful groups in the village ?
11. What kind of external connection do these powerful groups have?
12. Are there moneylenders and middlemen to collect fish from the village and advance loan ? Please check rate of interest, method/condition of loan and other obligations.
13. Is there an electricity supply in the village ? Yes/No. If yes, please check the frequency of power failure and duration failure on an average in a day/week.
14. What kind of transportation facilities exist in the village from the nearest market, city/town ?

	Type of Transport	Frequency in a day
1	*Pucca* road and state/private buses	
2	Private tempo/auto, bus/truck	
3	Only bullockcart and walk	
4	Bus - By walk to nearest road (write distance)	

15. What is the nearest railway station ? (Please write the name).
16. Are these transport facilities available all year round or only part of the year ? (Please describe).
17. What kind of educational facilities exist in the village of near the village ? (In case outside the village write name of the village and

distance from the village) :

(a) Primary school (for boys)
(b) Primary school (for girls)
(c) Primary school (mixed)
(d) Secondary school (for boys)
(e) Secondary school (for girls)
(f) Secondary school (mixed)
(g) ITI/vocational training/technical training
(h) Other (specify)

18. What kind of village level health facilities, including family planning centre, exist in the village ?
19. Where is the nearest hospital/doctor available ? (Write name of the place and distance).
20. Where do villagers go to market/sell and purchase of various domestic and agricultural related products :

(a) Name of the town/market
(b) Distance from the village
(c) Common mode of transportation
(d) Size of the market ?

21. Is there a weekly market near by :

(a) Day/weekly market
(b) Distance to weekly market
(c) Common mode of transportation ?

22. What kind of Panchayat Ghar (community hall) exists in the village ? (Describe).
23. Describe major landmarks and religious places in the village.
24. Are there wells/pumps for drinking water ?
25. Estimate the landholding size in term of :

(a) total agricultural land owned by the village (in acres) :

(i) irrigated
(ii) unirrigated

(b) small landholding size
(c) largest landholding size
(d) average landholding size

26. Are there any running co-operatives or organized, but not running ?
27. Write any other detail as seen important for the project.

ANNEX 3

Case Study of the Organization of Fishing Groups within a Village

Please actually describe the methods and techniques followed in :

(a) kin based fishing group; and
(b) non-kin based fishing groups.

The case study should cover :

(a) Who owns boat, nets etc. in the village for fishing ?
(b) How many people go fishing ?
(c) How the catch is divided amongst at fishing team members?
(d) How long has this practice has been followed ?
(e) What is the role of women in this kind of fishing ?
(f) How are the boat/nets and other equipment are maintained ? who pays the cost of maintenance ?
(g) What is the actual share that the individual fisherman gets from these fishing expeditions ?
(h) Are these variations in the system of distribution, catch and other methods among the villages and for different types of fish ?

***Note :** Please give all the details available and describe in detail.*

ANNEX 4

Group Level Interview Schedule Guide

(To be filled by two investigators simultaneously but separately during the group discussion)

Date/Starting Time

1. Name of the site : State
2. Name of the Village.
3. Name of the Taluka/Mandal/Block.
4. Name of the District.
5. Place/Location of the group meeting
6. Total number of estimated participants in the group discussion.

	Fisheries Caste/ Household	*Non-Fishing Caste/ Household*	*Village Lenders*	*Local Participants (write their no./names)*
Male Adults				
Female Adults				
Children				
Officials				
Total				

7. Describe the initial introduction about the proposed project and level of attention and participation of the villages.
8. Also describe the time taken and effort made in collecting the villagers.
9. Whether only a few tried to dominate the discussion or a large number of people participated in the discussion.
10. Did women participate in the discussion ?
11. Was there as need to stir the group discussion ?
12. What kind of an impression did you get about the whole group discussion ?
13. Are there any co-operatives in the village ?
14. What kind of co-operatives/individual arrangements are required by the people for running brackish water fisheries project.
15. Would you prefer to join a co-operative having your own caste mem-

bers, or would you like to have members from other castes also ?

16. Do you think people in your village are capable of running a fisheries co-operative on their own, or will they require help from outside? (Write details of the discussion).
17. Given the nature of the proposed "brackish water fisheries", do you think you can take care of your plot ? (If not, what kind of assistance do you think you will need?
18. Many of the villagers during the group discussion thought that running a co-operative with the help of the Fisheries Department and Government Officials is possible and desirable. What is your opinion ?
19. What kind of external assistance will be required for running the brackish water fisheries farm ?
20. Whom would you trust most for handling the money and affairs of the brackish water fisheries farm ?
21. Whom would you distrust most for handling the money and affairs of the brackish water fisheries farm ?
22. Are there people in your village who can work on the brakish water fisheries farm ? (Please write their names).
23. What kind of individual/group training the respondent thinks will be required to run the brackish water fisheries farm?
24. Would non-fishing families like to participate in ths project ?
25. How do you think women can help and participate it brackish water fisheries project ? (Probe for answer from both men and women respondents).
26. Suppose you are asked to take up a half acre plot for brackish water fisheries. What kind of help will you need ?
27. Can it be handled by your family members or will you need other members of the village to help you ?
28. Would you prefer to work independently on the half acre plot and take the risk of losing or earning more, or would you like to earn a fixed wage as an employee on the fish farm ?
29. Observe and describe the general lifestyle of the respondent and the level of conviction in his/her response.

Note : *More questions will be developed in this guide.*

ANNEX 5

Household Level Interview Schedule Guide : Respondent; Head of the Household

Date :

(Please fill in this section before starting the interview)

1. Name of the site : State
2. Name of the Village.
3. Name of the Taluka/Mandal/Block.
4. Name of the District.
5. Name of the respondent (Please give full name)
6. Sex. Male/Female
7. Age (in years)
8. Education : Illiterate/below primary level/below secondary level/ beyond High School (College) and above Technical/Vocational
9. Current occupation of the respondent.
10. Traditional occupation of the caste/household.
11. Religion. Caste
12. Whether SC/ST . Yes/No
13. House-type (observe, check and then record).
    Thatched hut/*kuchcha*/*pucca* having thatched roof/*pucca*/any other write.......
14. Land ownership.

    (a) Irrigated land (size in acres)
    (b) Unirrigated land (size in acres)

15. Cattle ownership :

    (a) Milch cow/buffalo (No.)
    (b) Any other domestic animal (No.)

16. Have you taken any loan in last three years ? Yes/No.
    If yes, amount of loan, purpose of loan, source of loan, method of repayment (write full details.)
17. Estimated annual total household income (in Rupees) *(Kindly probe carefully and collect data on all sources of income)*

18. Are you a member of any co-operative society ? Yes/No.
If yes, name of the co-operative society and where the respondent is an officer bearer. If no, have you ever applied to become a member? (Write all the reasons given).
19. Creating a hypothetical situation like "If you were asked to join the brackish water fisheries co-operative about which you heard earlier in the group meeting, would you like to join" ?
"Yes". Why ?
"No". Why ?
20. Check on mass-media exposure to know the level of external exposure. (Please describe and make judgement as to whether nil/poor/ high exposure.)
21. Would you prefer to join a co-operative having your own caste members or would you like to have members from other castes also ?
22. Do you think people in your village are capable of running a fisheries co-operative on their own, or will they require help from outside ? (Write details of the discussion).
23. Given the nature of the proposed "brackish water fisheries" do you think you can take care of your plot ? If no, what kind of assistance do you think you will need ?
24. Many of the villagers during the group discussion thought that the running of a co-operative with the help of the Fisheries Department and government officials is possible and desirable. What is your opinion ?
25. What kind of external assistance will be required for running the brackish water fisheries farm ?
26. Whom would you trust most for handling the money and affairs of the barckish water fisheries farm ?
27. Whom would you distrust most for handling the money and afffairs of the brackish water fisheries farm ?
28. Are there people in your village who can work on the brackish water fisheries farm ?
29. What kind of individual/group training does the respondent think will be required to run the brackish water fisheries farm ?
30. *Only to non-fishing respondent*
Having no previous experience in fishing of any kind, can you handle brackish water fisheries ? If yes, why would you like to change your present occupation ?
31. How do you think women can help and participate in brackish water fisheries project ? (Probe for answer from both men and women respondents).

32. Suppose you are asked to take up a half acre plot for brackish water fisheries, what kind of help will you need ?
33. Can it be handled by your family member, or will you need other members of the village to help you ?
34. Would you prefer to work independently on the half acre plot and take the risk of losing or earning more, or would you like to earn a fixed wage as an employee on the fish farm ?
35. Observe and describe the general lifestyle of the respondent and the level of conviction in his/her response.

Note : *More questions will be developed in this interview guide.*

APPENDIX - 2

QUERI - 1

Development and Educational Communication Unit, ISRO
SAC P.O. Ahmedabad - 300 053

Brackish Water and Inland Fisheries Project
Quick and Useful Evaluative Research Investigation (QUERI)

VILLAGE SCHEDULE GUIDE

1. Name of the Site. State

2. Name of the Village.

3. Name of the Taluka/Mandal/Block. District

4. Is this a permanenent/ semi-permanent seasonal settlement village?

6. What is the nearest Railway Station ? (please write the name and its distance from the village)

7. Is domestic electricity supply in the village ? Yes/No

8. Is Agriculture/Industrial electricity supply ? Yes/No
   If yes, please check note down the frequency of power failure and duration of failure on an average in a day.
   Now fill up the last page table

9. What kind of village level health facilities including family planning centre exists in the village ?

   (In case outside the village write name of the village and distance from the village).

10. What kind of transportation facilities exists in the village from the nearest market/city/town ?

Sr. No.	Type of Transport	Frequency in a day	All Weather/ Seasonal
1.	*Pucca* Raod - State/ Private buses		
2.	Tempo/auto/truck	X	
3.	Bullockcart/horsecart	X	
4.	By walk to nearest road (write distance)	X	

11. What kind of educational and Voccational/Technical Training facilities exist in the village or nearby ?

Educational Distance
1.. ....
2.. ....

Vocational/Technical
3..
....

12. What kind of leadership exists in the village
    (a) Traditional (caste based)
    (b) Traditional (village based)
    (c) Elected representative
    (d) Which caste dominates in Village Panchayat

13. Who are economically powerful groups in the village ?

14. What kind of external connection these powerful groups have ?

15. Are there middle men to collect fish from the village and advance

loan ? Please check rate of interest, method/condition of loan and other obligations.

16. Where do villagers go for marketing/sell and purchase of various domestic and agriculture product ?

— Name of the town/market Distance

— Common mode of transportation

17. Is there any weekly market nearby ?

— Name of the town/market Distance

18. What kind of Panchayat Ghar (community hall) exists in the village ? (Describe)

19. Are there wells/hand pumps for drinking water ? Yes/No

20. Estimate the iand holding size in terms of

(i) Total agricultural land own by the village (in acres)

(a) Irrigated
(b) Unirrigated

(ii) small land holding size
(iii) large land holding size
(iv) average land holding size

21. Are there any running co-operatives society or society which was organized but not running ?

22. Write any other detail as seen important for the project.

Village and Household Census

23. What is the present population of the village ?

24. What is religions/caste composition of the villages :

S. No.	Caste/ Religion	Total H.H.	Total Population	Main Occupation	No. of Fishermen	SES Poor	SES Mid.	SES Upper
1								
2								
3								
4								
5								
6								
7								
8								
9								
10								

HH = Household
SES = Socio-Economic Status

QUERI - 2

Development and Educational Communication Unit, ISRO
SAC P.O. Ahmedabad - 380 053

Brackish Water and Inland Fisheries Project

Quick and Useful Evaluative Research Investigation (QUERI)

Case Study of the Organisation of Fishing Groups within a Village

Please give all the details available and describe it

Please describe the methods and techniques of fishing followed in :

(a) kin based fishing groups, and
(b) non-kin based fishing groups.

The case study should cover :

(a) Who owns boat, nets etc., in the village for fishing ?

(b) How many people go for fishing ?

(c) Who takes the leading role ?

(d) How the supplies of fishing requirements are made ?

(e) What is the role of women in this kind of fishing ?

(f) How the boat/nets and other equipments are maintained ? Who shares the cost of the maintenance ?

(g) What is the actual share that the individual fisherman gets from these fishing expeditions ?

(h) Are there variations in the system of distribution catch and other methods among the village and for different types of fishes ?

(i) What are the training arragements ?

QUERI - 3

Development and Educational Communication Unit, ISRO
SAC P.O. Ahmedabad - 380 053

Brackish Water and Inland Fisheries Project

Quick and Useful Evaluative Research Investigation (QUERI)

GROUP LEVEL INTERVIEW SCHEDULE GUIDE

(To be filled by two investigators simultaneously but separately during the group discussion)

Date :
Starting time :

1. Name of the Site. State

2. Name of the Village.

3. Name of the Taluka/Mandal/Block.

4. Name of the District.

5. Place/Location of the group meeting

6. **Describe the init' introduction about the proposed project and level of attention and participation of the villages. Mention if there was a need to initiate group discussion**

7. **Also describe the time taken and efforts made in collecting the villagers. Also mention about the life style and the level of conviction in the respondents.**

8. Total number of estimated participants in the group discussion

	Fishing caste/ household	Non-Fishing caste/ household	Village leaders	Vocal participants (write their numbers / names)
Male Adults				
Female Adults				
Children				
Officials				
Total				

9. Whether only few tried to dominate the discussion or a large number of people participated in the discussion.

10. Did women participate in the discussion ?

12. What kind of impression you got about the whole group discussion?

Points for Group Discussion

13. Are there any co-operatives in the village ?

14. What kind of co-operative/individual arrangements are desirable by the people for running brackish water fisheries in the village ?

15. Will you prefer to join a co-operative having your own caste members or will you like to have members from other castes also ?

16. Do you think people in your village are capable of running fisheries co-operative on their own ? or they will require help from outside? (write details of the discussion).

17. Given the nature of proposed "brackish water fisheries" ? Do you think you can manage a plot (*use local term and usual size*)? If no, what kind of assistance you will need ? (Financial, managerial, and technical).

18. Many of the villagers during the group discussion thought that running of co-operative with the help of Fisheries Department and Government officials is possible and desirable. What is your opinion ?

19. What kind of external assistance will be required for running the brackish water fisheries farm ?

20. Whom will you trust most for handling the money and affairs of the brackish water fisheries farm ?

21. Are there people in your village who can work on the brackish water fisheries farm ?

22. What kind of individual/group training the respondent thinks will be required to run the brackish water fisheries farm ?

23. Will non-fishing families also like to participate in this project ?

24. How do you think women can help and participate in brackish water fisheries project ? (probe for answer to both men and women respondents)

25. Suppose you are asked to take up a one acre plot for brackish water fisheries : What kind of technical, managerial, and financial help you will need ?

26. Can it be handled by your family member ? or Will you need other members of the village to help you ?

27. Will you prefer to work independently on the 1/2 acre plot and take the risk of loosing or earning more or will you like to earn a fix wages as an employee on the fish farm ?

QUERI - 4

Development and Educational Communication Unit, ISRO
SAC P.O. Ahmedabad - 380 053

Brackish Water and Inland Fisheries Project
Quick and Useful Evaluative Research Investigation (QUERI)

Household Level Interview Schedule Guide : Respondent Head of the Household

(Please fill this section before starting the interview)

Date :

1. Name of the site. State
2. Name of the village. Taluka/Mandal/Block

....

5. Name of the respondent (please write full name)

....

6. Sex Caste (Mention if SC/ST) Religion

Male/Female ....

7. Education : Illiterate/below primary level/below secondary level/beyond higher secondary (college) and above/technical vocational.

8. Age Occupation of respondent

.................years ....

9. Traditional occupation of the caste/household

10. House-type (Observe, check and then record) thatched hut/*kuchcha*/*pucca* having thatched roof/*pucca*/any other write.

11. Landownership :

Irrigated land	size in acre	....
Unirrigated land	size in acre	....
Fish pond	size in acre	....

12. Cattle ownership : Milch Cow/Buffalow (number)

Any other domestic animal (number)

13. Have you taken any loan in last three years ? Yes/No

If yes, amount of loan, purpose of loan, source of loan, interest rate, repayment mode (write details).

14. Estimated annual total household income in rupees. *Kindly prove carefully and collect data on all sources of income.*

15. Are you a member of any co-operative society ? Yes/No

If yes, name of the co-operative society and whether the respondent is an office bearer ?

If no, have you ever been apporached to become a member of a co-operative society and why you did not become member ? (write the details)

16. Suppose if you were asked to join the Brackish Water Fisheries Co-operative about which you heard earlier in group meeting, will you like to join ?

"Yes", Why ?

"No" Why ?

17. Check on mass-media exposure to know the level of external exposure (please describe and make judgement whether it is nil/poor/high exposure).

18. Will you prefer to join a co-operative society having your own caste members or will you like to have members from other castes also ?
19. Do you think people in your village are capable of running a fisheries co-operative on their own ? or Will they require help from our side ? (write details of the discussion)
20. Given the nature of proposed "brackish water fisheries". Do you think you can take care of your plot ? If no, what kind of assistance do you think you will need ?
21. Many of the villagers during the group discussion thought that running of co-operative with the help of Fisheries Department and other Government Officials is possible and desirable ? What is your opinion?
22. What kind of external assistance will be required for running the brackish water fisheries farm ?
23. Can it be handled by your family member ? or Will you need other members of the village to help you ?
24. Will you prefer to work independently on the 1/2 acre plot and take the risk of loosing or earning more or will you like to earn a fix wages as an employee on the fish farm ?
25. Whom will you trust most for handling the money and affairs of the brackish water fisheries farm ?
26. Are there people in your village who can work on the brackish water fisheries farm ? (please write their names).
27. What kind of individual/group training you think will be required to run the brackish water fisheries farm ?
28. Will non-fishing families like to participate in this project ?
29. How do you think women can help an participate in brackish water fisheries project ? (probe for answer to both men and women respondents).
30. Suppose you are asked to take up a 1/2 acre plot for brackish water fisheries, what kind of help you will need ?
31. Observe and describe the general life style of the respondent and the level of conviction in his/her responses ?

APPENDIX - 3

IMPRESSION MANAGEMENT : A NOTE ON METHODS AND TECHNIQUES OF THE SURVEY

3.0 This short personal account of the Project Leaders has been documented to give some insights in to the field situation in which the survey was conducted. The note is intended to give detailed description which could not be provided in Chapter - 2 on research methods.

3.1 The Rapid Rural Appraisal (RRA) method has been in use for a long time by administrators, policy makers, and planners, for reviewing the situation. In the context of Brackish Water Fisheries Development Project, it was used in a slightly modified manner in the sense that we tried to quickly collect the "focused information" about the socio-economic life of the people around the proposed brackish water sites in a holistic perspective and also collected survey data. It has given a comprehensive picture of the situation. We had good experience of collecting data with the new technique which we had only used in television programme pretesting or need assessment studies.

Although, the sites were already selected, we were supposed to select the villages around the site in a radius of one to five km to meet the number of beneficiary households three times more than required. Before proceeding from for the field, we scanned through the secondary information available and made tentative circles on the maps to select survey villages. In this way, we prepared a list of probable villages.

But when we reached to the first site, the situation looked quite different and so we felt the need to modify our approach for the selection of the villages.

3.2 Some information about villages around these sites was also available at the fisheries directorate or assistant directorate levels. So the real picture emerged only when we reached the first site Jagatjore. So first we had to identify the location of the site and then plot all the villages around it and circle it again. In Jagatjore site, the villages have number of *Tolas* (wards). Each settlement is large enough to be called as village. The villages closer to the site were also the villages belonging to other *Panchayats* (Village Councils) or even Blocks. In C.G.Palem site, the term *Panchayat* is synonymous to an independent settlement village, whereas, the English revenue ward village is used for a cluster of such settlements.

These field realities, we started facing right from the selection of field investigators at Cuttack in case of Orissa and at Machilipatnam in case of Andhra Pradesh. Based on our previous experiences, we wanted to select four Masters degree holders in Social Sciences, half of whom should have been female investigators. We could not find enough number of such Social Scientists. Other problem which we faced was to hire the investigators from the district towns and to persuade them to stay with the team in the rural areas or site areas.

The research team tried to get government accomodation wherever possible. In most of the places not more than one room was made available for the entire team. In such a situation, the team had to stay in expensive (from their point of view) acomodation which created financial difficulties.

3.3 The field investigators hired could understand the local language and could explain and write the responses, observations and discussions in Hindi or English in the schedule guides .Due to language limitation on the part of the field investigators they had to try hard to explain many things with a very limited vocabulary. Same handicap was observed in filling up the schedules also in Andhra Pradesh. Initially, it was difficult to find out English speaking investigators. From Vijayawada, we tried to take the college students, but they were not ready to work in the village areas 100 km away from Vijayawada.

3.4 For our survey, we had approached the Directorate of Fisheries in the respective States through the Government of India and through them we contacted the district level functionaries. If facilitated our work, as we got all the possible data and other help from the officials. For Andhra Pradesh, one of the Hydrologists from the Department of Fisheries was deputed to coordinate the research team activities in the Machilipatnam and Ongole areas. But it created another kind of problem. This gentleman and the local fisheries inspectors took it as part of their prime responsibility to be with the research team. They helped the team to get the people collected and explain the project to the people. But because of their presence, the people probably, could not express their feelings about the fisheries department and in most of the cases they said that fisheries department will be very helpful in the brackish water shrimp farming. It is not that we did not contact the individual households without taking the fisheries inspector with us, but since we approached the villagers through the fisheries inspector; we feel that villagers had viewed us as one from the fisheries department.

3.5 In the field situation, we thought that we would be able to collect a large number of people and explain to them about the project, hold group discussion and to collect village level information like census. But when we reached the villages, we found it quite difficult to collect people for group discussion. In some of the villages, we found that Fisheries Extension Officer or the Fisheries Inspector were not acquainted with the villagers. So, it was difficult for them to collect people for group discussion. In many cases not more than 10-15 people would gather. During this survey period, the villagers were working in the field, women were attending to household work or were helping their men folk in the field or were working at the work places. Also, women did not openly participate in the discussion because of their social norms. The elderly women who came to see what we have been talking about, stood on the periphery of the group meeting and could answer only when persistently probed. However, in their homes, the women were vocal. Since many of them did not attend group discussion, we had to explain again to the respondent about the project. It took lot of time.

3.6 Jagatjore was the most difficult area from the communication point of view. Looking into the difficult approach road, it was decided that the team will camp in the village because it was not possible to come daily from the nearest town — Pattamundai, almost 20 km to conduct survey. The head master and the Panchayat Sarpanch (Head Village Council) were kind enough to arrange bicylces and one motor cylce to take the research team to the respective villages. Since we were totally dependent on the head master and the Sarpanch, we could not totally avoid their being with us and their more presence might have affected, to some extent, the quality of data.

3.7 Villages like Suniti from Jagatjore site and P.T.Palem village from

P.T.Palem site could not be approached. Sunity is four km and P.T. Palem is six km away from the last approach point. One has to cycle/ walk these distances. We could not arrange required number of bicycles for our team to go to these places. Riding bicycles on these *kuchcha* paths were not easay too. For these two villages, we collected all possible secondary data. Villagers of these villages were invited at the nearest point for discussion. About 10 people from Suniti came at Kantilo village. Three of the P.T. Palem villagers came to meet the research team in Machilipatnam. To reach to Kantilo of Jagatjore and Patapadu of C.G. Palem we used motorized boat. The motorized boat took us to Jagatjore market site, one and half km away from Kantilo. But we could to reach Pathapadu because of the low tide. The team had to walk for two and haft km to reach the village. The small creeks on the way was a real problem to cross. Motorized boat also took a long time (about an hour) to reach there.

Experiences/Problems of Group Discussion

3.8 After reaching the villages, first task was to collect the villagers at a place for group discussion. Different approaches were adopted at different places depending upon our mode of transport as well as the location of the village(s). In Jagatjore, when we reached near the villages of Sugal, Bandhapada, Gogua and Nanjura (all located at the Bund road side), we would stop at a place where we could find few villagers. Through the Sarpanch/head master/local leader/ward member, we introduced ourselves as well as the project very briefly. It was followed by a request to collect other villagers for group discussion. The team leader would emphasize that women should also be requested to participate in group discussion, although, it did not happen in all cases. In the villages of scattered population like Kalatunga and Guludia we reached the primary school and took help from the teachers to collect the people at schools, which were mostly located at a central place convenient for the villagers to come. In compact villages it took 15-20 minutes to collect the villagers but in scattered settlements, it took more than 30-45 minutes and we had to wait for the people to come. In Panikhia the ward member had come to Kalatunga and he knew that we would visit Panikhia so he collected five persons who were waiting for us when we reached the village.

In Kantilo, we had to reach by waterway so a message was sent to the villagers of Kantilo as well as to the villagers of Suniti to reach to the Primary School at Kantilo. When we reached the school, people of the village saw us walking and quite a few of them started gathering. Because of the scattered population, we could not collect more people. A few villagers from Suniti also came to Kantilo. It took 15-20 minutes for us to settle down.

3.9 On the first day, local officials, the village head or school teacher with the help of the Fisheries Officials explained about the project. Afterwards, one field investigator was made responsible to explain the project to the villagers. The introduction to the villagers took not more than 15 minutes. But Kantilo and Suniti villagers had more questions about the co-operation to be provided, the government guidelines and the loan. May be this was because they are more involved in co-operative societies and exposed to the bank at Suniti.

3.10 After an introduction to the project, and the research team, invariably the field investigators initiated the discussion. Opinion were sought about the project in general, before coming to the specific questions. Explaining to the villagers about the help/assistance required for the brackish water shrimp farming was extremely difficult task.

3.11 Similar approaches were follolwed every where including P.T. Palem and C.G. Palem site areas. In P.T. Palem, we mostly reached the villages by jeep except in case of K.P.T. Palem and Venkatadurgambapuram where we had to stop the jeep away from village and walked in the villages of Chintayagaripalem and Pedadevrampadu of C.G. Palem site. The team walked after crossing the river to Patapadu of the same site. It had to reach half way by motorized boat and then walk up to the villages. In these areas the fisheris inspectors were familiar to the area as well as to the key persons of the villages.

In P.T. Palem area, when the team reached K.P.T. Palem we had to get hold of a villager to collect the villlagers near the temple. In Malakayalanka, villagers became suspicious on seeing the government jeep. One of them came close and after knowing our purpose of visit he called others near the jeep. In Venkatadurgambapuram the Assistant Director (Fisheries), District Rural Developement Agencies (DRDA) had visited the village and left the village 15 minutes before we reached. So the villagers could see one jeep coming again and different people getting down and walking in the village. The prominent villgers and the office bearers of the fisherman co-operative society came to receive us. The villagers of Kammavaricheruvu knew about our visit of Malakayalanka and about the project on the previous day. So when we were going to Malakayalanka next morning three to four villagers stopped our jeep and requested us to visit their village also. Two of them also come to Malakayalanka when we were conducting our survey. When the team reached the village Kammavaricheruvu, there were 10-12 villagers gathered in the cyclone shelter to listen us.

The group discussion took almost 30 minutes to an hour. The group discussions were generally dominated by a few. The research investigators tried their best to take views of others present in the group discussion and encouraged them to express their views.

3.12 Village demographic data were collected from the most informed people in the village, but not necessarily the best informed one(s) all the time because of their being not available. Triangulation technique was used but could not help much since none of the respondents had accurate idea about the total population of the different castes. Often the informants found it extremely difficult to remember the accurate number of different castes staying in the village, particularly, in the large size villages. In small villages, however, they counted on the finger and tried to tell the number. In the single caste village or single caste majority village, we allowed the informants to count the number of minor castes and rest of the numbers were calculated as major castes. For villages with scaterred settlement like a few in Jagatjore area and the large size village like Vadapalem of P.T. Palem area or Gundayapalem of C.G.Palem, it was difficult to get individual household information from the respective settlements.

3.13 In order to make the triangulation technique more effective in the field situation, an effort was made to contact the village level workers. Getting reliable data was very difficult at Block level. In such a short visit, it was almost impossible to collect the data from each of the Gram Sevaks (village level workers). In some cases, wherever, data were obtained from Gram Sevak, it was quite difficult to use. *For example*, the statistics on population and land are based on District

Census 1981 and other data were based on surveys conducted in later years. The Gram Sevaks dit not remember the year of data collection. Therefore, it was not possible to recheck the data through triangulation technique.

3.14 Similarly when data on land, population, income and occupation were taken from two persons within the same village, the figures varied substantially, particularly, when the size of the village was large. Hence, triangulation technique has a limited utility for collecting reliable data on some of the variables described above in the context of village India.

3.15 The respondents had difficulty in answering the questions asked about the assistance they would require for individual plot or a farm. They did not know what to answer and what type of help they might require. Then wanted to know more about the project and other details from us. But even then, most of the time, they indicated the need for money and land, technical and financial assistance. Whenever, we tried to explain different kinds of help they might need, they only nodded their heads in affirmation. It was dificult to explain everything time and again.

3.16 We used our judgment in the purposive sample selection of the respondents.Geographical distribution of castes, economic classification and caste hierarchies were the basis of our selecting respondents for household level data collection from among the total population.

3.17 In every village, six representative individual heads of the households were supposed to be interviewed from among the people gathered for group discussion. This could not be followed in all the villages. In Jagatjore site area, we could interview 74 persons from 13 villages. In village Panikhia, we could interview only four persons because of the scattered settlement. For two other heads of households the field investigator tried but that being the market day, no respondent was available and the investigator could not contact them second time. The respondents of Suniti village were interviewed at Kantilo so we could get only four respondents representing four different caste groups of the village. In C.G. Palem site, the investigators had problem of getting enough respondents from a particular ward of village Chintayagaripalem so only four persons could be interviewed. In P.T.

Palem area when we were interviewing Malakayalanka villagers, two interviews were to be conducted by each of three investigators. But one of the investigators could collect data from one respondent only. As we were in hurry to leave the village before the sunset we could not interview the second person.

3.17 In village Dakshin Purshottampur, Dadanpatra site a special difficulty was faced in developing rapport with the villagers. Five hundred eighty-five acres of land was leased to Purba Purshottampur-Teghari Co-operative Society about 30-35 years ago. The adjoining Bengal Salt Company claimed 250 acres as their land and took possession forcibly from the Society. Almost all the villagers felt that some state officials were visiting this village frequently to persuade them to surrender the remaining land and Government then will develop brackish water ponds for poor people. The brackish water site is very close to this land. The villagers felt that we were also a part of the same government machinery, and it was very hard to convince that we were not at all concerned with the acquisition of their land.

3.18 In case of Panaspada, where majority of the villagers have unauthorised ponds in the Chilka area, did not reveal about the water body they possessed. At times, it was very difficult and time consuming to know the income generated by unauthorised fish catch. In Panaspada, such people are not one or two, but many.

Villagers' Hopes and Aspirations

3.19 Most of the fisherman and the poor villagers, looked upon us with some hopes that the project we were explaining will give them some economic benefits. But we faced resentment from the villagers of Chintayagaripalem in C.G. Palem area because the villagers wanted to use the proposed project site for salt farming and in the past this land was given to them for salt farming. Similarly, the Orissa Maritimes Chilka Area Development Agency (OMCADA) officials and labourers resented our visit because the proposed site for Jagatjore is presently under the control of OMCADA and the labourers fear loss of job if the site is taken away from the OMCADA. The Bengali women who are in the business of collecting prawn seeds were repeatedly asking questions if the project will develop hatcheries because of which their business might get negatively affected.

3.20 P.T.Palem area is under the influence of People's War Group (PWG) and the local officials insisted that the research team should leave the village before sunset. It had put a limitation on time and reaching very early in the morning in the village was not possible as we had to take our breakfast and carry out lunch for the day.

Field Problems

3.21 When the field investigators were asked to write a daily diary, they were reluctant to write it. They told us that whatever they had to be written, they had already written it in the schedule guides. Strict discipline is required in such a situation.

3.22 Checking all hundred per cent schedules after one comes back from the field late evenings was humanly impossible, particularly when each morning again the team had to go to the field. The time schedule for each site should be calculated after considering the condition of road and the time required to walk and cross waterways to improve the checking of the data.

APPENDIX - 4

LIST OF THE VILLAGES SURVEYED

State/Brackish Water Site		Name of Villages Visited (Location Code Number)	Name of Hamlets Visited*
West Bengal			
1. Dadanpatra			
	1	Purbbabar (R267) (Thakurchalk/Purushottampur Bard/ Puri Dal Bard)	
	2	Haurburi (R273)	
	3	Kismat Haurburi (R274)	
	4	Teghari (R275)	
	5	Purshottampur (R276) (Purba Purushottampur)	
	6	Dera (R280)	
	7	Sonamukhi (R282)	
	8	Purba Rania (R283) (Raniya/Rania)	
	9	Dadanpatra (R 284)	
	10	Dakshin Purushottampur (R286) (Purushottampur)	
	11	Machhlandapur (C334)	
2. Digha			
	1	Jhaugerya (R207)	
	2	Deulbatta (R208)	
	3	Gopalpur (R210)	
	4	Shankarpur (R224)	
	5	Maitrapur (D94)	
	6	Mirzapur (D95) (Purba Mirzapur)	
	7	Purba Mukundapur (D211) (Mukundapur)	
Orissa			
1. Jagatjore			
	1	Bandhapara (M11)	
	2	Kalatunga (M12)	
	3	Gogua (M13)	3.1 Ratanpur
	4	Baliganda (M14)	
	5	Dorigan (M15) (Doligan)	
	6	Nanjura (M16) (Lanjura)	
	7	Panikhia (M17)	
	8	Suniti (M37)	

	9 Kantilo (M39)	
	10 Sugal (M41)	
	11 Guludia (M42)	
	12 Kharianta (M43)	
2. Panaspada		
	1 Gopinathpur (B209)	
	2 Jadupur (B210)	
	3 Siara (B211) (Siara-Gola)	
	4 Panaspada (B212)	
	5 Sanabasantapur (B216) (Sanbasanpur)	
	6 Golapur (B217)	
	7 Rameshwarpur (B220) (Rameshwarpur/Pandapokhari)	
	8 Anandpur (225)	
	9 Gangadharpur (228)	
	10 Gorapur (229)	
Andhra Pradesh		
1. P. T. Palem	* 1. Palletummalapalem (56) (Bandar Taluk)	1.1 Kathatummalapalem 1.2 Kammavaricheruvu 1.3 Malakayalanka
	* 2. Pedayadara (21) (Divi Taluk)	2.1 Venkatadurgambapuram 2.2 Satyanarayanapuram 2.3 Vadapalem 2.4 Bhawanipurilanka
2. C. G. Palem		
	* 1. Devarampadu (8)	1.1 Gundayapalem 1.2 Chintayagaripalem 1.3 Pedadevarampadu 1.4 Sankatapallipalem 1.5 Harijanpadu 1.6 Pathapadu

! (R Code) = Ramnagar Police Station
! (C Code) = Contai Police Station
! (D Code) = Digha Police Station
! (M Code) = Mahakalpada Police Station
! (B Code) = Brahmgiri Police Station
* Revenue villages of the hamlet indicated against their numbers.

APPENDIX - 5

VILLAGE STATISTICAL PROFILE

Brackish Water Fisheries Site

Dadanpatra, Contai and Ramnagar Police Stations, Medinipur District, West Bengal, India

S. N	Name of the Village/ Hamlet	Approximate Distance to Contai Road Railway Station* in Km.	Domestic Electric Supply	Health Facility	Transport Facility in km.	Education Facility in the Village	Leadership	Dominant Caste	External Connection	Middlemen for Marketing Fish & Advance Loan	Market within Two km.	Panchayat Ghar (Village Council)	Co-operative Society
1	2	3	4	5	6	7	8	9	10	11	12	13	14
1.	Dakshin Purshottampur	17	No	No	Cycle rickshaw & waterway (km - 9)	None	Caste & village based	Other castes	None	No	No	No	Yes
2.	Macnhlandapur	18	No	No	Cycle rickshaw & walk (km - 10)	None	Village based	Other castes	Political	Yes	No	No	No
3.	Sonamukhi	23	No	No	Cycle rickshaw & walk (km - 2)	Primary School	Caste & village based	Other castes	Political & Business	Yes	No	No	No
4.	Purbbabar	20	No	No	Cycle rickshaw & walk (km - 5)	Primary School	Caste based & elected	None	Business	Yes	No	No	No

1	2	3	4	5	6	7	8	9	10	11	12	13	14
5.	Dadanpatra	21	No	No	Motorised	Primary School	Caste & village based	Scheduled castes	Business	Yes	No	No	No
6.	Purshottampur	22	No	No	Cycle rickshaw & walk (km. 7)	Primary School	Caste & village based	Other castes	Business	No	No	Yes	Yes
7.	Teghari	21	No	No	Cycle rickshaw & walk (km - 8)	Upper Primary School	Caste & village based	Scheduled castes	None	Yes	No	Yes	Yes
8.	Kismat Harburi	20	No	No	Cycle rickshaw & walk (km - 6)	Primary School	Village based	Scheduled castes	Political & Business	Yes	No	No	No
9.	Haurburi	21	No	No	Cycle rickshaw & walk (km - 7)	None	Village based	Other castes	Political & Business	Yes	Yes	No	No
10.	Dera	22	Yes	No	Motorised	Primary School	Village based	Other castes	Political & Business	Yes	No	No	Yes
11	Purba Rania	21	Yes	No	Motorised	Primary School	Village based	Other castes	Educational	Yes	No	No	No

* All distances from the nearest railway station have been approximated from the census data, information collected during the survey.

Brackish Water Fisheries Site
Dadanpatra, Contai and Ramnagar Police Stations, Medinipur District, West Bengal, India

Sr. No.	Name of the Village/Hamlet	Caste/Religion	Total Households	Total Population	Main Occupation	Socio-Economic Status		
						Lower	Middle	Upper
1	2	3	4	5	6	7	8	9
1.	Dakshin Purushottampur	Fishermen						
		SC/ST*	60	400	Fish.+Cult.		60	
		Other castes	240	1600	Fish.+Cult.		240	
2.	Machhlandapur	Fishermen						
		SC/ST*	10	75	Fishing		10	
		Other castes	45	325	Fish.+Other		45	
3.	Sonamukhi	Fishermen						
		SC/ST*						
		Other castes	150	900	Fishing		150	
4.	Purbbabar	Fishermen						
		SC/ST*	30	200	Fishing		30	
		Other castes	80	555	Fishing		80	
5.	Dadanpatra	Fishermen						
		SC/ST*	10	100	Fishing		10	
		Other castes	110	700	Fish.+Cult.		110	
6.	Purshottampur	Fishermen						
		SC/ST*	10	70	Fishing		10	
		Other castes	127	930	Fish.+Cult.		102	25

1	2	3	4	5	6	7	8	9
7.	Teghari	Fishermen						
		SC/ST*	40	300	Fish.+Labour		33	7
		Other castes	140	1120	Cultivation		90	50
8.	Kismat Haurburi	Fishermen						
		SC/ST*	5	30	Fishing		5	
		Other castes	30	370	Fishing		30	
9.	Haurburi	Fishermen						
		SC/ST*	5	55	Labour		5	
		Other castes	20	145	Labour		20	
10.	Dera	Fishermen						
		SC/ST*	75	700	Cult.+Labour		75	
		Other castes	175	1700	Cult.+Labour		175	
11.	Purba Rania	Fishermen						
		SC/ST*	35	300	Fishing		35	
		Other castes	40	300	Fish.+Service		40	
		Fishermen						
		SC/ST*	280	2230		273	7	
		Other castes	1157	8645		1082	75	
		Total	1437	10875		1355	82	

* SC/ST denotes Scheduled Castes/Scheduled Tribes.

Brackish Water Fisheries Site
Digha, Digha and Ramnagar Police Stations, Medinipur District, West Bengal, India

Sr. No.	Name of the Village/ Hamlet	Approximate Distance to Contai Road Railway Station* in km.	Domestic Electric Supply	Health Facility	Transport Facility in km.	Education Facility in the Village	Leadership	Dominant Caste	External Connection	Middlemen for Marketing Fish & Advance Loan	Market within Two km.	Panchayat Ghar (Village Council)	Co-operative Society
1	2	3	4	5	6	7	8	9	10	11	12	13	14
1.	Shankarpur	26	Yes	No	Motorised	Primary School	Village based	Other castes	Business	Yes	Yes	Yes	Yes
2.	Mirzapur	28	No	No	By walk (km-1.5)	Primary School	Panchayat based	Scheduled caste	Political	Yes	Yes	No	No
3.	Deulbatta	26	No	No	By walk (km.-2)	Primary School	Village based	Scheduled caste	Employment	Yes	Yes	No	No
4.	Purba Mukundapur	30	No	No	By walk (km.-3)	Primary School	Village based & elected	No data	No data	No	No data	No	No data
5.	Maitrapur	29	No	No	By walk (km.-2)	None	Village based	Other castes	Political, Government & Business	No	Yes	No	No
6.	Bopalpur	25	No	No	By walk (km-0.5)	Primary School	Village based	Other castes	Business	Yes	No	No	No
7.	Jhaugerya	27	No	No	By walk (km-3.5)	Primary School	Caste & village based	Scheduled caste	Business	Yes	No	No	Yes

* All distances from the nearest railway station have been approximated from the census data, information collected during the survey.

Brackish Water Fisheries Site
Digha and Ramnagar Police Stations, Medinipur District, Best Bengal, India

Sr. No.	Name of the Village/Hamlet	Caste/Religion	Total Households	Total Population	Main Occupation	Socio-Economic Status		
						Lower	Middle	Upper
1	2	3	4	5	6	7	8	9
1.	Shankarpur	Fishermen						
		SC/ST*	25	250	Fishing		25	
		Other castes	28	250	Fish.+Cult.		28	
2.	Mirzapur	Fishermen						
		SC/ST*	60	500	Fishing		60	
		Other castes						
3.	Deulbatta	Fishermen	25	300	Fishing		25	
		SC/ST*						
		Other castes						
4.	Purba Mukundapur	Fishermen						
		SC/ST*	70	350	Fishing		65	5
		Other castes	230	1125	Culti.+Other		218	12
5.	Maitrapur	Fishermen						
		SC/ST*	15	150	Fish.+Salt		15	
		Other castes	70	550	Fish.+Salt		70	
6.	Gopalpur	Fishermen						
		SC/ST*	36	200	Labour		36	
		Other castes	30	150	Labour		30	

1	2	3	4	5	6	7	8	9
7.	Jhaugerya	Fishermen						
		SC/ST*	40	400	Fishing		40	
		Other castes	30	300	Fishing		30	
		Fishermen	25	300			25	
		SC/ST*	246	1850			241	5
		Other castes	388	2375			376	12
		Total	659	4525			642	17

* SC/ST denotes Scheduled Castes/Scheduled Tribes.

Brackish Water Fisheries Site
Panaspada, Krishnaprasad Police Station, Puri District, Orissa, India

Sr. No.	Name of the Village/ Hamlet	Approximate Distance to Puri Railway Station* in km.	Domestic Electric Supply	Health Facility	Transport Facility in km.	Education Facility in the Village	Leadership	Dominant Caste	External Connection	Middlemen for Marketing Fish & Advance Loan	Market within Two km.	Panchayat Ghar (Village Council)	Co-operative Society
1	2	3	4	5	6	7	8	9	10	11	12	13	14
1.	Anandpur	43	Yes	No	By walk (km-2)	None	Elected	No data	None	Yes	Yes	No	No
2.	Gopinathpur	36	Yes	No	By walk (km-0.5)	High School	Elected & caste based	None	None	No	No	No	Yes
3.	Siara-Gola	38	Yes	No	By walk from Siara (km.-1.5)	Upper Primary School	Elected & caste based	Other castes	Political	Yes	No	No	Yes
4.	Gorapur	45	Yes	No	Waterway	Upper Primary & Middle English School	Elected	None	None	No	Yes	No	Yes
5.	Golapur	46	No	No	Waterway	None	No data	Other castes	Political	No	Yes	No	Yes
6.	Gangadharpur	46	Yes	No	By walk (km-2.5)	Lower Primary School	Caste based	No data	None	No	No	No	Yes
7.	Sanabasantapur	45	No	No	By walk (km-1.0)	U.G.M. School	Elected	None	None	No	No	No	Yes Unrecognised

1	2	3	4	5	6	7	8	9	10	11	12	13	14
8.	Panaspada	40	Yes	Yes	Motorised	College	Elected	None	None	Yes	Yes	Yes	Yes
9.	Jadupur	36	Yes	No	Motorised	High School	Elected & caste based	Other castes	None	No	Yes	No	Yes
10.	Pandapokhari	37	No	No	Motorised	Lower Primary School	Elected	Scheduled Caste	None	No	No	No	No

* All distances from the nearest railway station have been approximated from the census data, information collected during the survey.

Brackish Water Fisheries Site
Panaspada, Krishnaprasad Police Station, Puri District, Orissa, India

Sr. No.	Name of the Village/Hamlet	Caste/Religion	Total Households	Total Population	Main Occupation	Socio-Economic Status		
						Lower	Middle	Upper
1	2	3	4	5	6	7	8	9
1.	Anandpur	Fishermen						
		SC/ST*	20	90	Fish.+Cult.	15	5	
		Other castes	41	235	Fish.+Cult.	35	6	
2.	Gopinathpur	Fishermen	19	100	Fish.+Cult.	19		
		SC/ST*	66	385	Fishing	66		
		Other castes	94	600	Cultivation	94		
3.	Siara - Gola	Fishermen						
		SC/ST*	107	652	Fishing	70	27	10
		Other castes	53	400	Cultivation	32	21	
4.	Gorapur	Fishermen	30	200	Fishing	30		
		SC/ST*						
		Other castes	30	108	Cultivation	15	15	
5.	Golapur	Fishermen						
		SC/ST*	1	6	Fishing	1		
		Other castes	16	96	Fish.+Cult.	10	3	3
6.	Gangadharpur	Fishermen	70	800	Fishing	70		
		SC/ST*						
		Other castes						

1	2	3	4	5	6	7	8	9
7.	Sanabasantapur	Fishermen						
		SC/ST*	3	25	Cultivation	3		
		Other castes	74	675	Fish.+Cult.	74		
8.	Panaspada	Fishermen	90	475	Fishing		90	
		SC/ST*	7	30	Cult.+Fish.		7	
		Other castes	130	795	Cultivation		130	
9.	Jadupur	Fishermen	50	350	Fishing		50	
		SC/ST*						
		Other castes						
10.	Pandapokhari	Fishermen	40	150	Fishing	10	10	20
		SC/ST*						
		Other castes						
		Fishermen	299	2075		129	150	20
		SC/ST*	204	1188		155	39	10
		Other castes	438	2909		260	175	3
		Total	941	6172		544	364	33

* SC/ST denotes Scheduled Castes/Scheduled Tribes.

Brackish Water Fisheries Site
Jagatjore, Mahakalpada Police Station, Cuttack District, Orissa, India

Sr. No.	Name of the Village/ Hamlet	Approximate Distance to Puri Railway Station* in km.	Domestic Electric Supply	Health Facility	Transport Facility in km.	Education Facility in the Village	Leadership	Dominant Caste	External Connection	Middlemen for Marketing Fish & Advance Loan	Market within Two km.	Panchayat Ghar (Village Council)	Co-operative Society
1	2	3	4	5	6	7	8	9	10	11	12	13	14
1.	Gogua	116	Yes	No	By walk (km - 17)	Middle English School & U.P.M.E. School	Village based	Other castes	None	No	No	No	No
2.	Bandhapada	116	No	No	By walk or bicycle (km - 18)	Non-formal Primary School	Elected	Other castes	None	No	Yes	No	No
3.	Baliganda	118	Yes	No	By walk (km - 19)	None	Elected	No data	None	No	No	No	No
4.	Doligan	122	No	No	By walk (km - 23)	Upper Primary School	Elected & caste based	No data	None	Yes	No data	No	Yes
5.	Kalatunga	123	No	No	By walk (km - 4)	Middle English School	Elected	Scheduled Caste	None	No	Yes	No	No
6.	Guludia	122	Yes	No	By walk (km - 23)	Middle English School	Elected	No data	None	Yes	Yes	No	No

1	2	3	4	5	6	7	8	9	10	11	12	13	14
7.	Sugal	118	No	No	By walk (km - 21)	None	Elected	No caste based dominance	None	No	Yes	No	Yes
8.	Nanjura	125	No	No	By walk (km - 26)	Upper Primary School	Elected	Other castes	None	Yes	Yes	No	Yes
9.	Kharianta	117	No	No	By walk (km - 20)	None	Elected & caste based	Other castes	None	No	Yes	No	Yes inactive
10.	Panikhia	124	No	No	By walk (km - 25)	None	Elected	None	None	No	Yes	No	No
11.	Kantilo	125	No	No	By walk (km - 26)	Lower Primary School (private)	Elected	No data	None	No	Yes	No	Yes
12.	Suniti	127	No	No	By walk (km - 28)	Lower Primary School unrecognised	Village based & elected	None	None	No	Yes	No	Yes
13.	Ratanpur	117	No	No	By walk (km - 18)	Middle English School	Village based & elected	None	None	No	Yes	No	Yes

* All distances from the nearest railway station have been approximated from the census data, information collected during the survey.

Brackish Water Fisheries Site
Jagatjore, Mahakalpada Block, Cuttack District, Orissa, India

Sr. No.	Name of the Village/Hamlet	Caste/Religion	Total Households	Total Population	Main Occupation	Socio-Economic Status		
						Lower	Middle	Upper
1	2	3	4	5	6	7	8	9
1.	Gogua	Fishermen						
		SC/ST*	58	300	Labour	58		
		Other castes	103	700	Cultivation	81	20	2
2.	Bandhapada	Fishermen						
		SC/ST*						
		Other castes	36	230	Cultivation	8	18	10
3.	Baliganda	Fishermen						
		SC/ST*	3	12	Wax	3		
		Other castes	43	224	Cultivation	43		
4.	Doligan	Fishermen						
		SC/ST*						
		Other castes	53	300	Cultivation	30	23	
5.	Kalatunga	Fishermen	30	200	Fishing	30		
		SC/ST*	30	260	Cult.& Labour	27	3	
		Other castes	42	262	Cult.+Misc.	36	6	
6.	Guludia	Fishermen	50	300	Fish.+Cult.	44	6	
		SC/ST*	4	13	Cultivation	2	2	
		Other castes	8	104	Cultivation	7	1	
7.	Sugal	Fishermen						
		SC/ST*						
		Other castes	88	777	Cultivation	38	39	11

1	2	3	4	5	6	7	8	9
8.	Kharianta	Fishermen						
		SC/ST*	17	73	Cult.+Labour	17		
		Other castes	55	267	Cultivation	55		
9.	Nanjura	Fishermen						
		SC/ST*	30	117	Fish.+Wash	30		
		Other castes	116	621	Cultivation	116		
10.	Panikhia	Fishermen						
		SC/ST*	36	200	Fishing	36		
		Other castes						
11.	Kantilo	Fishermen						
		SC/ST*	42	230	Weave+Labour	42		
		Other castes	53	230	Cultil.+Misc.	52	1	
12.	Suniti	Fishermen	10	40	Fishing	10		
		SC/ST*	210	900	Weave+Labour	210		
		Other castes	220	1135	Cultivation	200	20	
13.	Ratanpur	Fishermen	6	20	Fishing	6		
		SC/ST*	9	40	Labour	9		
		Other castes	62	340	Cultivation		62	
		Fishermen	96	560		90	6	
		SC/ST*	439	2145		434	5	
		Other castes	879	5190		666	190	23
		Total	1414	7895		1190		

* SC/ST denotes Scheduled Castes/Scheduled Tribes.

Brackish Water Fisheries Site

P.T. Palem, Bandar and Divi Taluk, Krishna District, Andhra Pradesh, India

Sr. No.	Name of the Village/ Hamlet	Approximate Distance to Machlipatnam Railway Station* in km.	Domestic Electric Supply	Health Facility	Transport Facility in km.	Education Facility in the Village	Leadership	Dominant Caste	External Connection	Middlemen for Marketing Fish & Advance Loan	Market within Two km.	Panchayat Ghar (Village Council)	Co-operative Society
1	2	3	4	5	6	7	8	9	10	11	12	13	14
1.	Venkata-durgmbapuram	22	Yes	No	By walk (km - 2)	Primary School	Elected caste & village based	None	CASA Organisation	Yes	No	No	No data
2.	Satyanarayana-puram	22	Yes	No	Motorised	Primary School	Elected	Other castes	Political	No	No	No	Yes
3.	Vadapalem	21	Yes	No	Motorised	Primary School	Elected	Other castes	Political	No	No	No	Yes
4.	Bhavanipuri Lanka	22	No	No	Motorised	None	Village based	Other castes	CASA organi-sation	No	No	No	Yes
5.	Kathatummala Palem	31	Yes	No	By walk	Lower Primary School	Village based & elected	None	Fishermen Panchayat in P.T. Palem	Yes	No	No	No
6.	Kampavari-cheruvu	30	Yes	No	Motorised	Primary School	Elected	Other castes	None	No	No	No	Yes
7.	Malakayalanka	32	No	No	By walk	Lower Primary School	No data	None	PWG	No	No	No	No

* All distances from the nearest railway station have been approximated from the census data, information collected during the survey.

Brackish Water Fisheries Site

P.T. Palem, Bandar and Divi Taluk, Krishna District, Andhra Pradesh, India

Sr. No.	Name of the Village/Hamlet	Caste/Religion	Total Households	Total Population	Main Occupation	Socio-Economic Status		
						Lower	Middle	Upper
1	2	3	4	5	6	7	8	9
1.	Venkata-durgambapuram	Fishermen	162	700	Fishing		162	
		SC/ST*						
		Other castes	3	16	Misc.		3	
2.	Satyanarayanapuram	Fishermen						
		SC/ST*	127	488	Cult.+Labour	120	5	2
		Other castes	89	305	Cultivation	84	4	1
3.	Vadapalem	Fishermen						
		SC/ST*	26	157	Labour	26		
		Other castes	279	1234	Cultivation	179	50	50
4.	Bhavanipuri Lanka	Fishermen						
		SC/ST*						
		Other castes	30	200	Cultivation	20	10	
5.	Kathaummala Palem	Fishermen	116	500	Fishing	116		
		SC/ST*						
		Other castes						
6.	Kammavaricheruvu	Fishermen						
		SC/ST*	2	9	Labour	2		
		Other castes	63	271	Cult.+Labour	40	20	3

1	2	3	4	5	6	7	8	9
7.	Malakayalanka	Fishermen	38	150	Fishing	38		
		SC/ST*						
		Other castes						
		Fishermen	316	1350		316		
		SC/ST	155	654		148	5	2
		Other castes	464	2026		326	84	54
		Total	935	4030		790	89	56

* SC/ST denotes Scheduled Castes/Scheduled Tribes.

Brackish Water Fisheries Site

C.G. Palem, Ongole Taluk, Prakasam District, Andhra Pradesh, India

Sr. No.	Name of the Village/ Hamlet	Approximate Distance to Karavadi Railway Station* in km.	Domestic Electric Supply	Health Facility	Transport Facility in km.	Education Facility in the Village	Leadership	Dominant Caste	External Connection	Middlemen for Marketing Fish & Advance Loan	Market within Two km.	Panchayat Ghar (Village Council)	Co-operative Society
1	2	3	4	5	6	7	8	9	10	11	12	13	14
1.	Gumdayapalem	9	Yes	No	Motorised	Upper Primary School	Elected	Other castes	No data	Yes	No	No	Yes
2.	Chintayagiri Palem	11	Yes	No	By walk (km - 3)	Primary School	Caste based	Other castes	None	Yes	No	No	Yes
3.	Pada Devarampadu	11	Yes	No	By walk (km - 3)	Primary School	No data	Other castes	None	No	No	No	No
4.	Sankatapalli Palem	7	Yes	No data	Motorised	Lower Primary School	Elected	Other castes	None	No	No	No	No
5.	Harijanpadu	7	Yes	No	Motorised	Primary School	Elected	Scheduled castes	None	No	No	No	No
6.	Devrampadu	7	Yes	No	Motorised	Primary School	Elected	Other castes	Political	No	No	Yes	No
7.	Pathapadu	12	Yes	No	By walk (km - 7)	Lower Primary School	Elected	Other castes	Political	No	No	No	Yes

* All distances from the nearest railway station have been approximated from the census data, information collected during the survey.

Brackish Water Fisheries Site
C.G. Palem, Ongole Taluk, Prakasam District, Andhra Pradesh, India

Sr. No.	Name of the Village/Hamlet	Caste/Religion	Total Households	Total Population	Main Occupation	Socio-Economic Status Lower	Middle	Upper
1	2	3	4	5	6	7	8	9
1.	Gundayapalem	Fishermen	700	3000	Fishing		700	
		SC/ST*						
		Other castes	12	58	Salt		12	
2.	Chintayagari Palem	Fishermen	143	703	Fish.+Labour		143	
		SC/ST*						
		Other castes						
3.	Padadevarampadu	Fishermen						
		SC/ST*						
		Other castes	70	280	Labour		70	
4.	Sankatapalli Palem	Fishermen	50	300	Fish.+Labour		50	
		SC/ST*	10	35	Labour		10	
		Other castes	20	90	Washing		20	
5.	Harijanpadu	Fishermen						
		SC/ST*	220	800	Labour		220	
		Other castes						
6.	Devarampadu	Fishermen						
		SC/ST*						
		Other castes	100	800	Cultivation		75	25

1	2	3	4	5	6	7	8	9
7.	Pathapadu	Fishermen						
		SC/ST*	1	6	Labour		1	
		Other castes	117	609	Salt+Labour		115	2
		Fishermen	893	4003			893	
		SC/ST	231	841			231	
		Other castes	319	1837			292	27
		Total	1443	6681			1416	27

* SC/ST denotes Scheduled Castes/Scheduled Tribes.

APPENDIX - 6

CASE STUDIES OF CO-OPERATIVE SOCIETIES

Sidheswari Co-operative Society — Gogua : A Case Study from Jagatjore Brackish Water Site

Village Gogua is situated in Mahakalpada Block of District Cuttack, Orissa State. The village is situated at a crow distance of two km from Babar and the road distance is eight km from Jagatjore site. There is a *kuchcha* road which connects this village with Babar, a large village having marketing facilities. The wooden bridge connecting this village from Babar is very high and therefore, except motor cylces no other vehicle can reach this village. Another communication link is through the Jagatjore Brackish Water Site which is also *kuchcha.* The village is connected to Jagatjore by two road links. One of the shorter links is via Sugal. The condition of these roads is equally bad. No automobile can run on this road. Even driving a motorcycle is very risky. The second link road which is a little longer is also like the former one. The people of this village come to Babar for making day to day purchase. Village Gogua has one small shop which keeps some of the necessary provisions and *pan* (betel nut leaf). There is no health facility. There are some handpumps which are generally out of order and so people go to open wells to fetch drinking water. The nearest post office is located in Chokadi Gogua, which is located at a distance from Gogua. The village has a primary school. Another public building is of that the revenue inspector. The Gram Panchayat headquarters is located at Chokadi Tojha.

There are a total of 161 households, out of which 103 belong to the other castes category and remaining 58 are Scheduled Castes households. The entire village population is Hindu by religion except one Muslim family.

There are four houses which have a *pucca* roof. The remaining are thatched roof *kuchcha* houses. There is no electricity and some poles have been erected for future electrification.

Looking at the caste composition, it is found that people belonging to other castes are engaged in agriculture while the Scheduled Castes work as agricultural labourers. It is quite interesting to know while the men of the Scheduled Castes are generally engaged in agriculture, women of their community are engaged in fishing, and a little in raising ducks are poultry. Yet another important point to note is that most of the households of the Scheduled Castes possess agricultural land. Until 1970, most of the area of this village was under mangroves. However, with the passage of time and population pressure, the area was converted into agricultural land. While 70 per cent of the land was traditionally under cultivation, the remaining 30 per cent was mangroves and belonged to the State Government. It is very important to understand the process of distribution of the 30 per cent mangroves land among the villagers. It is from this point that the existing so called co-operatives society has come up. The people of this village started cultivating the Government land without waiting for any official distribution of the land by the Government. Having cultivated the land for some years, the Government then provided them lease on permanent basis.

The Sidheswari Co-operative Society is an unregistred society and is engaged in the fishing. Their site of fishing is located very close to Jagatjore. The society was organized in 1981 with 12 office bearers. Eleven of these officials belong to the village Gogua and one belongs to the adjoining village Baliganda. All these people belong to Scheduled Castes Kandara Caste. Three persons of this village, one of whom is now a ward member of the Panchayat, belong to this village. These three persons took the initial lead and took nine more people with them to form the co-operative. These 12 people encroched the land of the revenue and forest department. The total land of the society at the site near Jagatjore area which was under cultivation was about 450 acres and they had a total membership of 272 persons. The understanding between the members and the officials remained unwritten. Every year the office bearers enter an agreement with the private financiers and take Rs. 10,000-20,000 for strengthening the dykes and similar other works. They are under an obligation to sell their produce to the financier finally. There is no effort of using any improved seed for cultivation. The fish produce is screened and the prawn is sold to the financer. The remaining fish is sold in the local market. The cashier keeps the account and distributes the income to the members.

In the year 1987, the forest department forced thes people for evacuation of the land. So they were left with 111 acres. The *Tahsildar* of this area also issued evacuation notices to these people and charged Rs. 13,000 as penalty. The charges were unsually high because the category of the land was for the revenue of *fish Saurat*. In 1988, the fishery department tried to encourage the society and now they have submitted their papers for registration as co-operative society to the fisheries department. Presumably since the title of the land owned by the society is under encumbrance, the registration of the society has not taken place.

From this analysis, it is clear that this is the only co-operative society in this Gram Panchayat and Block. The three persons quite intelligently understood the process of distribution of Government agricultural land. The land was first unauthorised cultivated by the people of this village and later on was regularised by a permanent lease by the Government. The society has been paying heavy penalty and yet continuing because these three officials of the society have felt that sooner or later the presently unauthorised land of the revenue department will be allotted to them. They have also felt that these twelve officials of the society will not able to put pressure on the government hence, they included many more members. The distribution of the presently occupied 111 acres land among 286 members is perhaps the most difficult task for the State Government. Equal distribution will provide on an average water body of 111/286 acres, which is uneconomical for tiger prawn cultivation. Dividing the water body into 1/2 ha plot would mean its availability to only so many persons. In the absence of any formal record kept by the society it would be difficult to distribute the land if so needed.

For the past two years, the members of the society stay in their villages and the entire area is looked after by only one person who is the caretaker living on the site. This person is responsible for catch and all other aspects and the sale of the products is done by these three influential office bearers of the society. They take advance from the financier and sell the fish to the financier and utilize the money for penalty, strengthening of dykes and also for the legal litigations. It is understood that the society has not able to earn anything yet they are continuing with the hope that the Government will kindly allot the land to them.

Maa Ramachandi Primary Prawn Culture and Marketing Co-operative Society Limited, Kholanai : A Case Study from Jagatjore Brackish Water Site

This is the most recent co-operative society organized in Jagatjore Brackish Water Site. The details of the society are provided here. It was not possible to get more details as the society has just started functioning.

Date of Registration : 29 October 1990 (Regn. No. 317)

Governing Body

President of the society : Fisheries Extension Officer, Mahakalapada Block

Chief Extension Officer : Member of Governing body

Members

Name	Age	Occupation	Education	Caste Category
Magunicharan Ront	50	Fishing	Below Primary	SC
Keshab Chandra Bhol	26	Fishing	Above Primary	Other Castes
Shaneswar Ront	23	Fishing	I.S.C.	Other Castes
Dettep Bhol	21	Fishing	Under Matric	Other Castes
Bhaskar Nath	52	Fishing	Below Primary	Scheduled Castes (Tanti)
Secretary/Organizer —	Maheswar Ront — Interviewer Age : 30 Education : Matriculation Ocupation : Fishing Caste : Other Castes (Pradhan)			

2. Members : 22 from 16 villages : Koda Nai, Badini, Akadpuri, Kakatpur, Badabindha, Sanabaincha, Ranki, Bijayanagar, Arada, Maliancha, Kanpur, Sunniti, Kandara, Patia, Kharianta, Vateni, Kholanai. These members belong to fishermen as well as other castes group.
3. Total allotted land : 11 hectares
4. Site of Operation : Kakadpur and Arada and Akanpur' village in Mahakalpada Block.
   Distance from Jagatjore : 8-15 km
   Distance from Babar : 1-1/2 km

Sampradaya Salt Manufacturing Co-operative Society : A Case Study from C. G. Palem Brackish Water Site

The villagers of this area have been taking benefit of the high salinity of the ground water and doing salt farming on their individual plots. In August, 1990, almost 300 villagers of Pathapadu, Devarampadu, Gundayapalem and Chintayagaripalem have formed a co-operative society namely "Sampradaya Salt Manufacturing Co-operative Society". They are manufacturing salt of their own fields and trying to get government land for collective farming.

The individual landowner or with the held of hired labourers (locally known as *Koolies*) till the land. The plots of 0.25 acre is supplied with enough water by bore well to make muddy soil. This muddy soil is mixed using feet and 10-20 people work together to repeat this process for a month alternatively in 10 such plots. Mud soil levelled in one plot is allowed to dry by the time other is being prepared. After preparing a fully solid layer of levelled soil they allow it to dry and after almost 15 days they fill it with water and the water is allowed to dry for 15 days. The deposited soil with high salt content is taken out and stored at a central place near the field. This process is repeated as long they get bright sun till the middle of June.

Farmers manufacture salt individually or in a group of 3-4 famrers, mostly friends coming together. This raw salt is put together in co-operative form and by hiring a boat from the fellow villager it is taken to Motmala about 3 km from the village for selling it to salt agents. From Motmala, it is transported to Hyderabad for processing. Bullockcarts are also used for transporting salt from individual plots to the market place. The salt measurement unit is on per bullockcart basis and the price they get is shared accordingly after deducting the hiring charge of boat/cart. The salt manufacturers get a net benefit of approximately Rs. 2,500 per ha.

Almost 150 acres of land (100 acres towards eastern part and 50 acres on western part of canal) is under salt farming. The Salt Commission of Andhra Pradesh has approached the revenue department to give 150 acre of land on lease for salt farming.

Sri Rama Fishermen Co-operative Society : A Case Study from P. T. Palem Brackish Water Site

Sixty members of the village Venkatadurgambapuram formed a fishermen co-operative society in 1974. In course of time, the number of members has swollen to 168. In the beginning, the membership fee was only Rs. 5 which has increased to Rs. 10 by now. Like any other fishermen co-operative society, this society is also registered with the Department of Fisheries and the department conducts election every three years for the post of President, Secretary and Executive Committee Members. The president keeps account of the society. The major activity and contribution of the society is to get loan for buying the boat and getting nets on subsidy. The Andhra Pradesh branch of National Co-operative Development Corporation gives Rs. 3,000, subsidy for all the boats and 33 per cent subsidy (up to maximum of Rs. 1,000) per member on the nylon yarns purchased through co-operative societies. The usual cost of boat of 21 feet or 28 feet is Rs. 15,000 to Rs. 20,000 which they have to repay in a monthly installment of Rs. 500. For Beach Landing Craft (BLP) there is a soft loan of Rs. 50,000 having a low interest rate.

So far, the society has acquired 12 country boats and two BLPs, the fibre made motorized boat. The boats are used for fishing in Krishna River and in the Bay of Bengal. The selection of beneficiaries for giving boat is done on the basis of their interest for fishing, their skill of fishing, and their past history of payment of any loan taken.

The boat owners generally go for fishing in Krishna River with their family members. Also they go with other villagers (mostly close relatives or friends) for fishing at the bay. The catch is shared among those who go for fishing. Twenty per cent share is kept aside as boat charge. Generally, catch is sold to the agent. In most cases, only good quality fish is accepted by the agent. The fish not bought by the agent are distributed among the members who generally consume it or dry it for future use and sale. Sometimes the women go to sell the fish in villages. The non-fishing families do not have to buy fish because they can help the group in cleaning the net, carrying the fish load and get some small size fish for their consumption.

To select the beneficiary for a boat, the society has prepared a check list for future distribution of boat when purchased. The BLPs do not have subsidy on it so there is not much demand for it.

INDEX